Fluid Catalytic Cracking with Zeolite Catalysts

CHEMICAL INDUSTRIES

A Series of Reference Books and Text Books

Consulting Editor

HEINZ HEINEMANN

Lawrence Berkeley Laboratory
University of California
Berkeley, California

Volume 1: Fluid Catalytic Cracking with Zeolite Catalysts, *Paul B. Venuto and E. Thomas Habib, Jr.*

Additional Volumes in Preparation

Fluid Catalytic Cracking with Zeolite Catalysts

PAUL B. VENUTO
and E. THOMAS HABIB, JR.
Mobil Research and Development Corporation
Field Research Laboratory
Dallas, Texas

MARCEL DEKKER, INC.
New York and Basel

Library of Congress Cataloging in Publication Data

Venuto, Paul B. [Date]
Fluid catalytic cracking with zeolite catalysts.

(Chemical industries ; v. 1)
Includes index.
1. Catalytic cracking. 2. Zeolites. I. Habib, E. Thomas, [Date] joint author. II. Title. III. Series.
TP690.4.V46 665'.533 79-11529
ISBN 0-8247-6870-1

This monograph was originally published as an article in CATALYSIS REVIEWS—SCIENCE AND ENGINEERING, Volume 18, Number 1, 1978.

Marcel Dekker, Inc.
270 Madison Avenue, New York, New York 10016

Current printing (last digit):
10 9 8 7 6 5 4 3 2 1

Printed in the United States of America

PREFACE

Chemists and engineers quickly discover that fluid catalytic cracking is a very demanding field. It is difficult to be effective while working on only one aspect of catalytic cracking, and ultimately one will have to cope with the other dimensions of what is really an extraordinarily complex process. This book represents an attempt to cover the broad scope of fluid catalytic cracking, with emphasis on the highly-coupled interactions of the process — between the feedstock, the catalyst, the process hardware, and the desired products.

The discussion centers around the use of the zeolite cracking catalysts, but refers to the older amorphous variety where pertinent. Although fluid catalytic cracking (FCC) is the dominant theme, reference to bead catalysts is made where it is useful, i.e., as in discussions of diffusion limitations. The book also provides an integration of older technology (including the earlier reviews of the subject) with the modern, placing in juxtaposition the details of catalyst structure, chemical mechanism and kinetics, the hardware, and the context of a modern commercial petroleum refinery. The geochemical origins of petroleum are briefly outlined, because this provides a unique insight into the generation of cracking stock impurities, their fate in downstream processing, and their ultimate effect on the environment.

We hope that, because of its breadth and relatively small volume, this book will be useful not only to those involved in petroleum refining research, but also to those interested in catalysis in general, or in petrochemicals, whether they are in industry, at a university, or in government.

Paul B. Venuto
E. Thomas Habib, Jr.

CONTENTS

Fluid Catalytic Cracking with Zeolite Catalysts

I. INTRODUCTION

Fossil fuels, amassed over eons of geologic prehistory, are being irreversibly depleted about 10^6 times more rapidly than they were formed. With this sort of kinetics, there is little question that there exists, in the ultimate sense, an energy crisis [1-3]. That is, there will be fuel deficits for whatever time interval declining volumes of hydrocarbon reserves are not compensated for by increased supply from new energy sources such as nuclear, solar, and geothermal. In the interim period, the petroleum refiner is forced to utilize higher molecular weight, more hydrogen-deficient crudes, and crude fractions which have higher concentrations of impurities. Eventually, substantial quantities of "synthetic crudes" derived from coal, oil shale, tar sands, and bitumens may also become available.

There will be no slackening in demand for premium liquid fuels such as gasoline and distillate in the foreseeable future [1]. Roughly 50% of a barrel of crude refined in the United States is processed to gasoline [4] (Fig. 1), and this is largely at the expense of heavy fuel oil. Overseas demand for gasoline has traditionally been satisfied by naphtha boiling range fractions naturally present in crude [4] (Fig. 1). However, as crude oil supplies decline and prices rise, increased European demand for gasoline and light distillate will be met by upgrading residual fuel oil. This tendency will be greatly enhanced if coal, gas, or nuclear energy displace heavy fuel oil in power plants and related applications. The continuing importance of and demand for a flexible molecular weight-reducing process—catalytic cracking—seems a reasonable expectation. This is typified by the 5.6 million bbl/day of catalytic cracking capacity operative in the United States in 1975 [6]. The importance of catalytic cracking is indisputable [7], and the economic impact of zeolite catalysts on catalytic cracking is tremendous. Over 90% of United States catalytic cracking units were employing zeolite catalysts by 1969 [8]. An economic study published in 1966 estimated that the savings to refiners alone, from use of zeolite cracking catalysts over the older amorphous forms, were \$250,000,000/yr [9].

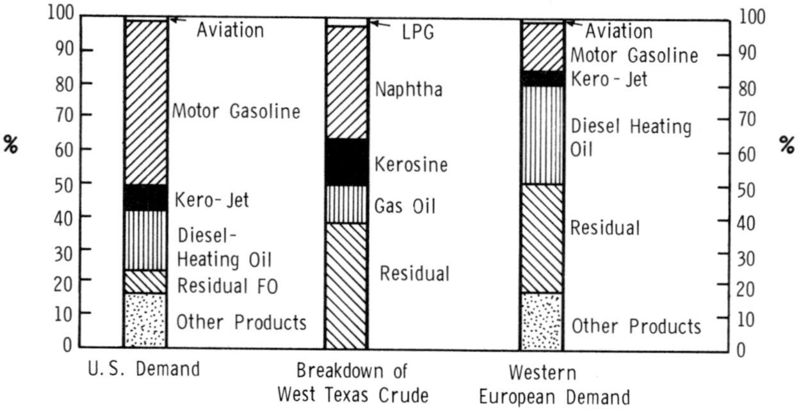

FIG. 1. Comparison of refined product demand in the United States and Western Europe vs composition of a "typical" crude; adapted, in part, from Refs. 4 and 5.

II. SCOPE

Reviews by Sittig [10] (1952) and Voge [11] (1958) provide excellent background on early engineering and catalyst work in catalytic cracking. Pohlenz [12] described operational variables in fluid catalytic cracking (FCC) in a short review, while Hatch [8] highlighted chemical aspects of cracking in the refining environment. More recently, Oblad [13] has focused on zeolite cracking catalysts. Salient historical features in the history of catalytic cracking are summarized in Table 1.

The operation of a large, commercial fluid catalytic cracker involves a multiplicity of complex interdependent interactions (Fig. 2). In the present review we attempt to represent (although not exhaustively) the state-of-the-art of catalytic cracking. We will cast this in the frame of reference of these interactions—between the catalyst on the one hand and the restraints of the feed (nature), the product configuration, and the process unit (engineering/hardware) on the other. Discussion will largely be in terms of fluid cracking, since this mode is most commonly employed in modern refineries, although moving bed (TCC) systems and bead catalysts will be discussed where appropriate.

III. GENESIS AND COMPOSITION OF PETROLEUM: IMPLICATIONS FOR DOWNSTREAM CATALYTIC PROCESSING

A. Geochemical Origins of Crude

Many of the catalyst and engineering restraints observed in catalytic cracking processes can be related to the presence of certain nonhydrocarbon impurities in crude, and it is instructive to trace their origins. There is convincing evidence that the preponderance of petroleum known today is derived from ancient biological material [21-23], particularly from lipid-rich lower marine plants such as plankton [22]. Much of this occurred in the Mesozoic and Tertiary periods, geologic eras in which the greatest abundance of potential hydrocarbon-producing primitive life flourished. This enormously complex process is visualized in Fig. 3, which clearly associates the existence in petroleum of heteroatoms such as nitrogen, oxygen, and sulfur with their presence in the biochemical building blocks—lipids, proteins, carbohydrates, and the like—of earlier life forms. Highly specific confirmation for this hypothesis can be found by studying biological markers [21], "chemical fossils," the structure of which can be traced to previous biological sources.

Examples of direct interest to cracking are the metalloporphyrins found in petroleum [24-27], in particular the nickel and vanadium forms, which are notorious poisons for cracking catalysts [12, 28, 29]. One porphyrin structure dominant in plants is that of chlorophyll [24] (Fig. 4), but this contains magnesium rather than heavy metals. However, much evidence points to a biogenic origin for petroleum porphyrins, and, for the case of chlorophyll, it is not

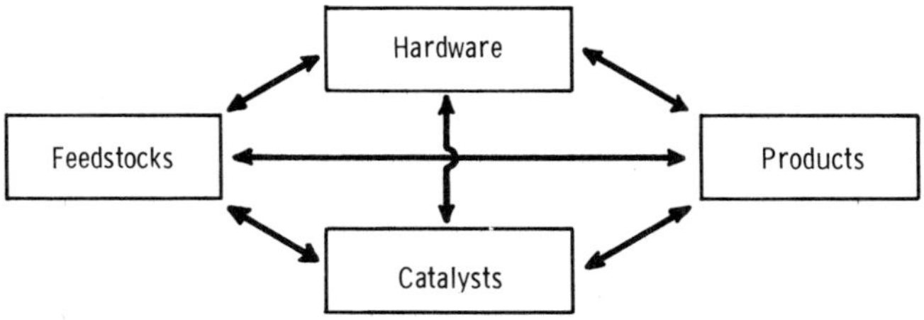

FIG. 2. Interactions in fluid catalytic cracking: FCC involves an incredibly complex multidimensional interplay of variables.

TABLE 1

SOME IMPORTANT EVENTS RELATING TO CRACKING

Event	Year of commercialization or implementation	Significance	Refs.
Thermal cracking	1912	Initial major molecular weight reducing process	8
Tetraethyl lead	1923	Octane appreciator for cracked gasoline; over 50 yr useful life	14
Catalytic polymerization	1936	Recombination of olefinic cracked fragments to gasoline	8
Catalytic cracking	1936	Greatly improved efficiency with use of natural clays (montmorillonite) as catalyst	10
First commercial silica/alumina produced	1940	Synthetic SiO_2/Al_2O_3 with superior properties became dominant cracking catalyst	13
World War II	1941-1945	Great demand for aviation gasoline and other liquid fuels; spurred acceleration of refining process development	
Original upflow FCC	1941	Model I, pioneer fluid cracker	10

Thermofor catalytic cracking (TCC)	1942	First moving bed bead catalyst cracking process	10
Paraffin alkylation	1942	Efficient process for combining cracked fragments to high octane gasoline	8
Synthesis and commercialization of zeolite X	1954	Future component of zeolite cracking catalysts	15, 16
Synthesis and commercialization of zeolite Y	1959	Future component of zeolite cracking catalysts	15
Hydrocracking	1960	Cracking of heavy feeds in presence of hydrogen and dual functional catalyst; growth was slowed upon advent of zeolite catalysts	8
Zeolite cracking catalysts	1962	Zeolite incorporated into matrix; revolutionized catalytic cracking	9, 13 17, 18
Short contact time riser FCC	1971	Maximum utilization of active zeolite catalysts, minimum overcracking and coke formation	11, 19
Controlled combustion zeolite cracking catalysts	1975	Promoted zeolite cracking catalysts capable of converting CO to CO_2; improved yields, selectivity; lower CO stack emissions	20

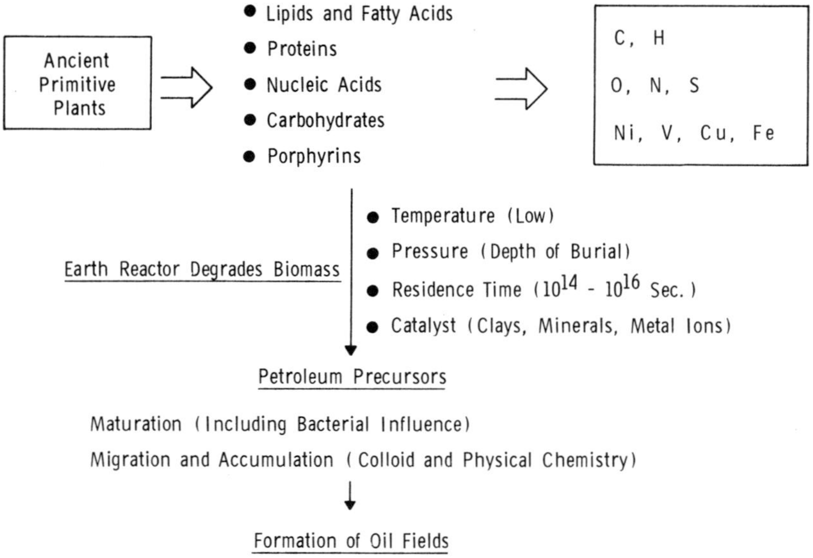

FIG. 3. Visualization of origin of petroleum from ancient biological reactants.

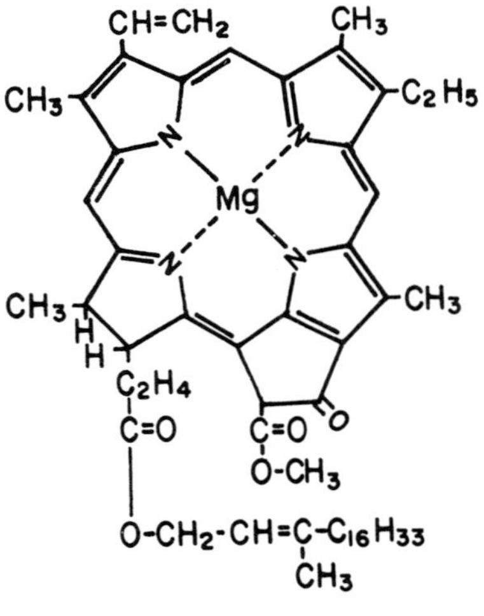

FIG. 4. Structure of chlorophyll a: Ni, Cu, Fe, and V form chemically-resistant complexes with tetradentate, highly conjugated porphyrin ring systems available after Mg-extrusion.

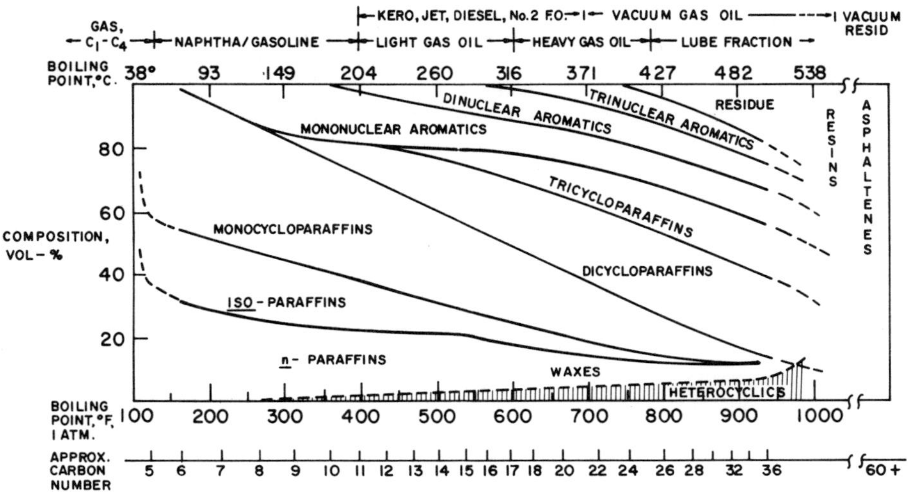

FIG. 5. Relative quantities and boiling range of the major hydrocarbon classes in the crude from the Ponca City Field, Ponca City, Oklahoma; adapted from Refs. 31-33.

difficult to postulate a sequence starting with magnesium cleavage where heavy metals complex with pigment degradation products at later stages [24, 25]. The isolation from crude oils of pristane and phytane [30], isoprenoid paraffins that could have derived from the phytyl side chain of chlorophyll, complements this hypothesis. As expected, both thermodynamic and chemical (acid resistance) stability of metalloporphyrins increase in the order V > Ni > Mg [25].

B. Composition of a Typical Crude and Implications for Processing of FCC Feeds

1. Hydrocarbons

In Fig. 5 the relative quantities of the major hydrocarbon classes found in a "typical" mid-continent-type crude [31, 32] are plotted vs the boiling range in which they occur. This will serve as a frame of reference for later discussion of feed-catalyst interaction in catalytic cracking. The major classes of hydrocarbons in crude are paraffins, naphthenes or cycloparaffins, and aromatics; olefins are not generally present. The number of carbon atoms in petroleum hydrocarbons ranges from C_1 to C_4 (small amounts of dissolved gases) through C_{60} as a maximum [34], although the heteroatom-containing resin/asphaltene fractions may include moieties with molecular weights in the thousands. At C-numbers of 18 and higher, the n-paraffins may be classified as waxes. Material from almost every boiling range of crude has been catalytically cracked in one application or another.

Note from Fig. 5 that the relative fraction of aromatics increases with boiling point, while that of paraffins decreases; in fact, at about 950° F, almost 50% of the hydrocarbon is aromatic, and over 80% of it contains rings. Also, the high boiling range fractions contain large proportions of condensed polynuclear aromatics. Homologs of phenanthrene are far more abundant than those of anthracene, and chrysene homologs predominate over pyrene [34]. Note also the increase in hydrogen deficiency as feed aromaticity increases.

Asphaltenes are heterocyclic substances usually containing sulfur, oxygen, nitrogen, and metals, and possessing molecular weights in the range of 900 to 3000 (or greater) [34]. Geochemically, asphaltenes are most likely secondary products derived by condensation and polymerization of heavy organic debris during diagenesis [34]. The structure of asphaltenes conceived by Dickie and Yen [35] is represented in Fig. 6; these are colloidal in nature [26, 36], generally soluble in benzene but insoluble in C_3 to C_7 paraffins [34, 36], and have relatively loose linkages between different groups of condensed rings. A large fraction of the polynuclear aromatics in asphaltenes is peri-condensed, as in anthracene [37]. Some micelles may be as large as 100 to 300 Å in diameter [35]. The C/H ratios of asphaltenes may range from ~0.77 to 0.95 (7 to 9% H) [37]. The term "resin" includes polar materials of different structures which tend to be strongly adsorbed on the surfaces of polar solids; these have some degree of solubility in light paraffins and have molecular weights on the order of 500 to 1000 [34]. When adsorbed on cracking catalysts, resin/asphaltene materials are potent coke precursors [38, 39].

2. Impurities

Table 2 lists the levels of sulfur, nitrogen, and metallic impurities found in various crude oil sources; also shown are values for carbon residue, which is to some extent related to coke-forming tendencies [39, 41]. These are "average" or "typical" values assembled from various literature sources [40, 42] and serve to illustrate the wide variations in these contaminants. Their presence places many significant restraints on catalytic cracking, and necessitates implementation of considerable peripheral "cleanup" technology. Note the

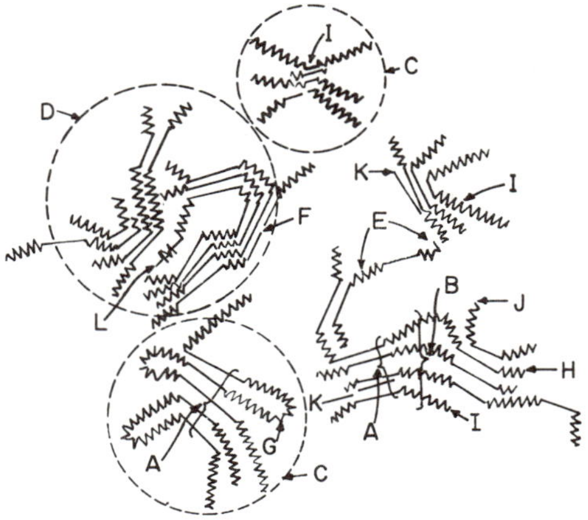

FIG. 6. The macrostructure of asphaltenes is essentially a three-dimensional network of condensed aromatic rings, short aliphatic chains, and naphthenic structures. (A) Crystallite. (B) Chain bundle. (C) Particle. (D) Micelle. (E) Weak link. (F) Gap and Hole. (G) Intracluster. (H) Intercluster. (I) Resin. (J) Single layer. (K) Petroporphyrin. (L) Metal. After Dickie and Yen [35].

relatively minor quantities of S, N, O-type heterocyclics in the boiling range 275 to 980° F superimposed on the n-paraffin region of the Ponca City crude [33] Fig. 5). The quantities of heteroatoms in many crudes (Wilmington, California [33], for instance) increase substantially at boiling points greater than 500° F. Note the very high sulfur levels—with their ultimate environmental impact—in Boscan and most of the Middle East crudes (Table 2).

Nitrogen compounds in petroleum are mainly concentrated in heavy fractions and residue (which must eventually be cracked), particularly in resin/asphaltene units [34, 43]. Their levels vary from less than 0.1% in East Texas or Burbank crude (corresponding to a few percent of N-compounds) to as high as 0.66% in Wilmington crude (representing more than 10% of N-compounds). Basic nitrogen compounds, i.e., pyridines (C_6 to C_{10}), quinolines (C_{10} to C_{17}), and their hydro-derivatives, comprise 25 to 35% of the total nitrogenous material [34, 43]; these strongly absorb at acidic catalyst sites and act as poisons [44-47]. Carbazoles, indoles, pyrroles, etc., constitute the neutral fraction. The ratio of basic to nonbasic nitrogen compounds is greatly affected by hydrotreating, i.e., of shale oil fractions [48, 49].

TABLE 2

Typical Impurity and Carbon Residue Levels of Some Representative Crudes [40]

Crude source	Gravity, °API	Sulfur, wt%	Nitrogen, wt%	Metals, ppm		Carbon residue, wt%
				Ni	V	
East Texas	37.6	0.31	0.07	3	4	2.4
West Texas	37.6	1.67	0.08	5	10	1.7
Oklahoma (Burbank)	38.6	0.27	0.05	-	-	0.9
Wyoming (Little Mo)	23.3	2.90	0.24	16	83	8.9
California (Wilmington)[a]	19.5	1.37	0.66	51	53	6.3
Alaska (Prudhoe Bay)	27.0	1.07	0.23	11	31	4.7
Venezuela (Boscan)	11.3	5.62	0.61	112	819	5.2
Nigeria (East Imo River)	24.0	0.31	0.16	9	4	1.9
Libya (Amal)	36.2	0.27	0.09	-	-	2.9
Kuwait	31.7	2.47	0.15	7	43	5.1
Arab Light (South Arab Fields)	34.4	1.73	0.11	4	16	2.4
Safaniya (Bahrain)	27.7	3.04	0.17	20	52	4.0
Iran (Gach Saran)	29.3	1.72	0.23	33	123	4.6
Indonesia (Minas)	32.5	0.08	0.11	7	-	2.9

[a] Los Angeles Basin.

A range of about 20 to 200 ppm of metals is typical of most of the world's crude, although levels in United States heavy oil deposits, Canadian tar sands, and the Venezuelan Orinoco Belt may range from 300 to 1500 ppm Ni and V [50]. Nearly half the metals in the periodic table have been identified in crude oil, but the most abundant heavy metals are V, Ni, Fe, and Cu [43]; V and Ni levels for a representative group of crudes are listed in Table 2. Although traces of metals are occasionally found in gas oil fractions (largely as oil-soluble porphyrins), the concentration of metals steadily increases with molecular weight of distillate fractions, being maximum in the asphaltene fraction [26, 34]. Within the asphaltene fractions, some metals are in porphyrin complexes (see Fig. 6, K) [34], while others appear to be held in coordination complexes with heteroatom elements of asphaltenes (such as sulfur compounds) [26]. A significant fraction of the iron present can be from corrosion of pipes, vessels, etc. ("tramp" iron). Due to the association of many oil reservoirs with brines, appreciable quantities of Na and Ca salts may be present dissolved in emulsified water, hence the need for desalting [51]. A sample of North Slope crude was reported to contain 10 lb of salt/1000 bbl [52]. Entrained silt, sediment, and even heavy ions such as barium from drilling muds may occasionally be present.

IV. THE PLACE OF THE CATALYTIC CRACKER IN A TYPICAL REFINERY SCHEME

In a sense it is justified to refer to a fluid catalytic cracker (FCC) as the heart of a modern United States refinery oriented toward maximum production of gasoline. The cracker operation ultimately affects units that fractionate, generate, or clean up feeds upstream to the FCC, and also interacts with processes downstream that likewise separate, clean up, and further process its broad spectrum of products. There is thus a high degree of almost homeostatic interdependency among the units in such a refinery complex, which still permits great flexibility in optimization of products. Various refinery configurations that include crackers have been reported, some of which include cokers [5, 53-56], hydrocrackers [5, 54, 56], solvent deasphalting [57, 58], and feed hydrotreaters [53, 57-59], and all of which require unsaturated gas plants and alkylation. Figure 7 is a schematic of a cracking-alkylation configuration which combines features reported by Skamser [58] and Olson and Hutchinson [56], and shows several different approaches to handling of vacuum residua; in this case, some of FCC heavy cycle oil is hydrotreated prior to recycle. Atmospheric tower bottoms could also

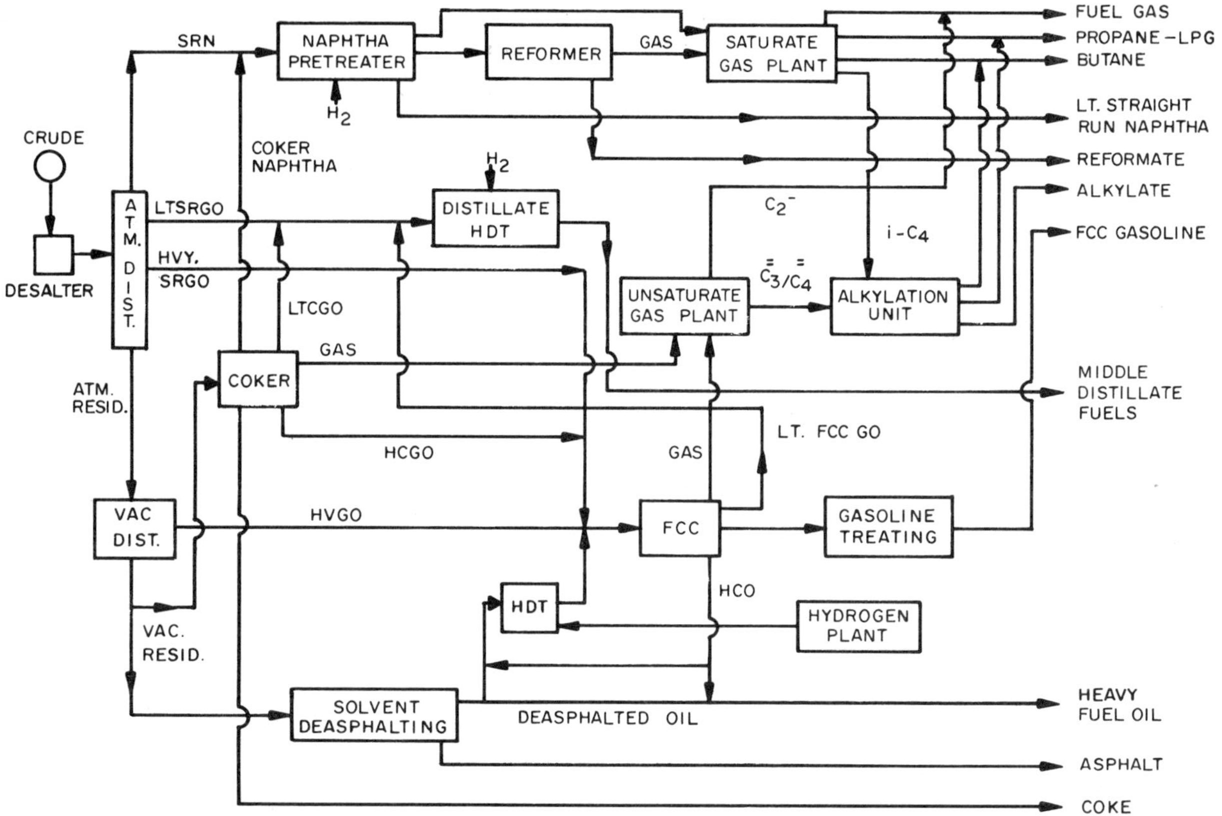

FIG. 7. Conceptual gasoline-oriented refinery featuring catalytic cracking and alkylation (modified from Refs. 56 and 58).

be routed directly to the FCC [60]. Alternatively, heavy coker gas oil and/or heavy vacuum gas oil could be hydrotreated and cracked. Stoichiometric balances for several refining schemes have been described [53, 56, 58, 60]. Almost any crude fraction can be fed to an FCC, but before describing these in detail we must first describe the unit itself.

V. FCC COMPLEX AND HARDWARE

Numerous designs of FCC units have appeared in the recent literature [12, 61-66]. Generally the FCC complex may be broken down into three major sections [67]: (a) reactor-riser, (b) regenerator-flue gas handling, and (c) fractionation. Associated with the cracker, but commonly outside the battery limits, are (d) an unsaturated gas plant and (e) an alkylation unit. In Fig. 8 a "typical" modern FCC complex is depicted, representing a composite of reported features [12, 19, 61, 66, 68]. The numbers (1)-(22) in Fig. 8 refer to feed/product streams for the complex mass balance.

A. Reactor-Riser Section

Fresh feed (1) after residence in the feed drum to allow settling of dense entrained material is passed through a series of feed/effluent heat exchangers, allowing stepwise heating of feed and simultaneous cooling of product streams. Fresh feed may be joined by heavy cycle oil (HCO) (2) and slurry settler bottoms (3) recycle streams [12, 61, 66], and then may be passed through a (fired) feed preheater [10, 12, 69]. The combined feed (5) is then contacted with hot regenerated catalyst (6) at the base of the riser, the feed is vaporized, and the mixture of hot vapor and catalyst lifted—with simultaneous catalytic reaction—up the vertical riser. Most conversion occurs in the riser [19, 61, 64, 65, 70], with the reactor serving as separator of catalyst dust and hydrocarbon vapor [70, 71]. Proper design provides near plug flow of catalyst and oil, thus minimizing back-mixing of gas with spent catalyst [64].

Features of a riser-separator/reactor zone adopted for an FCC process with short catalyst/oil contact times are detailed in Fig. 9. After disengagement from catalyst, hot product vapors (7) are sent to fractionation. Spent catalyst (8) is passed through a baffled stripping zone countercurrent to a flow of steam [62, 65, 68, 72, 73], where

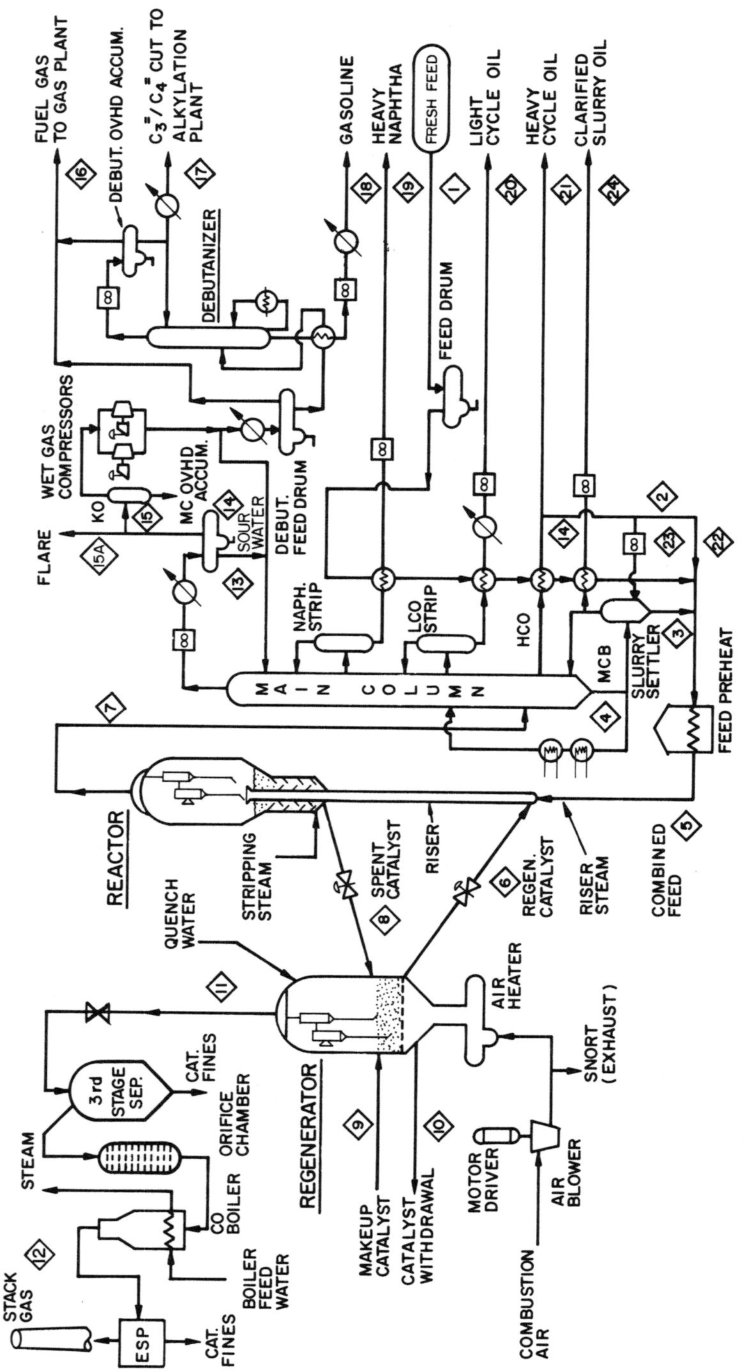

FIG. 8. Modern short contact time riser FCC complex.

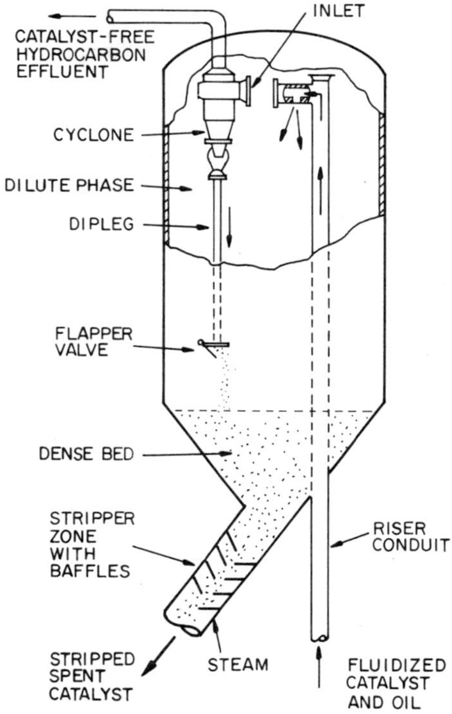

FIG. 9. By allowing riser to discharge downward toward dense bed and close to inlet of cyclone separation, undue agitation of dense bed—with consequent secondary reactions—is avoided; dipleg may also extend below interface between dense and dilute phases. After Conner, Strother, and Vermilion [70].

entrained, hydrogen-rich hydrocarbon vapors [12, 69, 72] are displaced upward prior to flow of coked catalyst to the regenerator. The stripper does not remove adsorbed hydrocarbons [12, 69], but displaces material present in the interstices between catalyst particles. There is an optimum steam rate [12, 62, 65, 74], above which no beneficial effects occur, and below which the regenerator temperature may rise rapidly. Proper control of residence time in the stripper is necessary if excess air consumption in regenerator is to be avoided [74-76]. Reactor pressure is maintained by pressure control of the main column receiver [12], and is closely coupled to fractionator/gas plant operation.

B. Regenerator and Flue Gas Circuit

The coked catalyst (8) passes through the spent catalyst slide valve (operated by a catalyst level controller in the reactor [12]) and then enters the regenerator. Catalyst flow into the regenerator may be tangential [62] as well as downward. The major purpose of regeneration is to oxidize the coke on the spent catalyst with oxygen (to form CO, CO_2, and H_2O), thereby reexposing the catalytically-active surface of the fluid particles [62, 65, 74, 77-79]. After compression, combustion air enters the bottom of the regenerator through a grid distribution pattern designed to provide efficient mixing [62, 78] of air with the coked catalyst. This generates a fluidized dense bed [80] in which superficial velocities may be as high as 3.4 ft/sec [62]. Regenerated catalyst leaves via a standpipe which is circumferentially remote from the spent catalyst inlet in order to avoid "short circuiting" [62]. The nearly coke-free catalyst, after passing through the regenerated catalyst slide valve (operated by the reactor temperature controller [12]), is then admixed with oil at the riser base and the cycle is repeated. Provisions are made for adding fresh catalyst makeup to maintain inventory and for withdrawal of aged or contaminated catalyst [62]. Carbon on regenerated catalyst averaged 0.25 wt% in 1976 (vs 0.33 wt% in 1970) [81].

Gaseous combustion products and some entrained catalyst are conveyed upward out of the regenerator dense bed into a dilute phase zone, where sets of primary and secondary cyclones [73, 82-84] remove and return the entrained catalyst and route the flue gas through a plenum chamber into the overhead flue gas circuit. Figures 10 and 11 depict details of a typical regenerator vessel [85, 86]. Csicsery et al. [87] have reported catalyst and flue gas sampling techniques wherein the course of regeneration can be monitored during operation.

From the plenum, the flue gas (with much reduced catalyst loading) may pass to a third stage separator [77, 88], where essentially complete removal of particles greater than 5 μm is typically achieved. After passing through the flue gas slide valve and orifice chamber (which together reduce the flue gas to atmospheric pressure [88]), the flue gas may be routed to a CO-boiler [73, 78, 88]. Here, CO can be completely converted to CO_2 to meet emission standards and to recover the resulting heat of combustion and sensible heat of the flue gas in high-pressure steam. There is also the option of incorporating a power recovery system into the flue gas/air blower system [88, 89]. Before venting through the stack (12), the flue gas is sometimes passed through electrostatic precipitation (ESP) [57, 88].

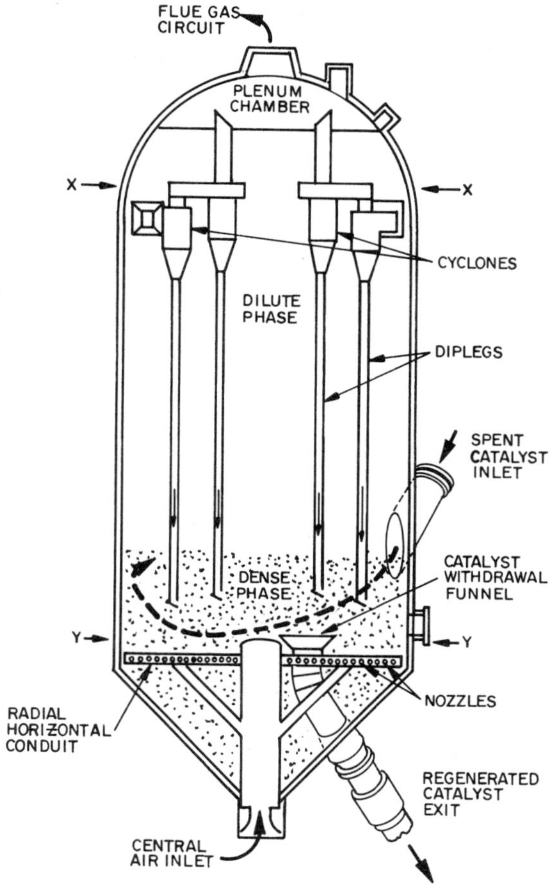

FIG. 10. Schematic view of regenerator with tangential catalyst inlet and bed circulating circumferentially about vessel vertical axis; ΔP over nozzles may be in the range of 0.4 to 1.2 psi to keep all the nozzles active. Adapted from Refs. 85 and 86.

An instrument sensitive to the ΔP between reactor and regenerator vessels controls the regenerator pressure by operation of the flue gas slide valve [12]. Flow rates of most major feed and product streams are usually measured via orifice meters [90]. Various aspects of the historical evolution of FCC units from their commercialization in the early 1940s through modern designs have been discussed in the literature [10, 19, 61, 66, 91, 92].

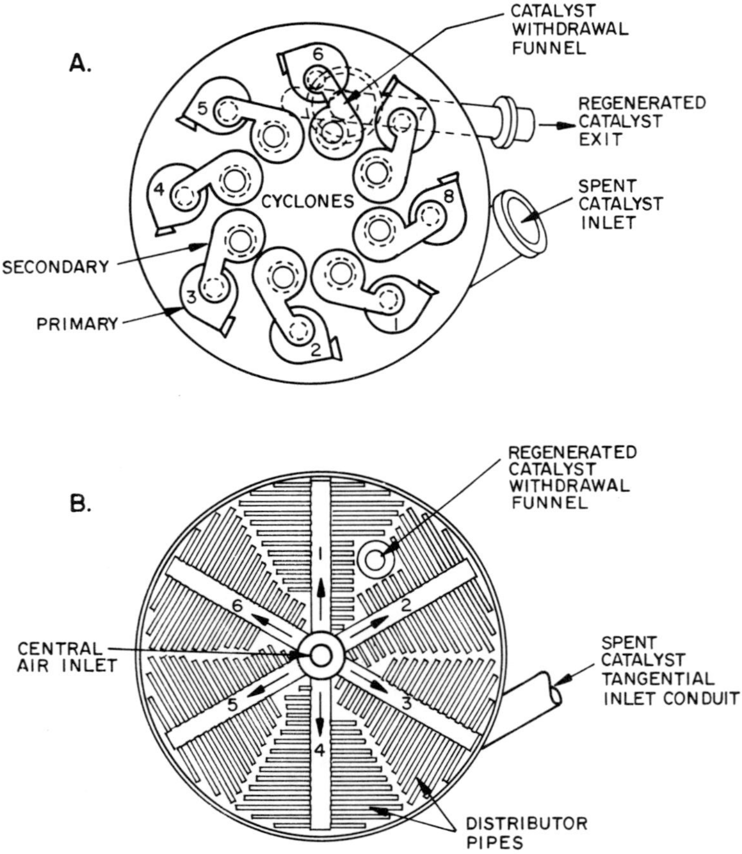

FIG. 11. Schematic cross-sectional views (A) through X-X in Fig. 10, showing eight primary and secondary cyclones for removal of catalyst entrained in flue gas; (b) through Y-Y in Fig. 10, showing segmented regeneration gas-distributing grid comprising six radiating, horizontal, closed-end conduits (1-6) which emanate from a central regenerator air/gas inlet. From each of the radial conduits a plurality of spaced, horizontal distributor pipes issues, each of which possesses numerous nozzles (see Fig. 10) for air distribution. Adapted from Refs. 85 and 86.

C. Fractionation System

Reactor effluent (7) is fed into the base of the main column where, with heat removal at various elevations [12, 74], it is fractionated into selected product streams. Gasoline and lighter are taken overhead to gas concentration units [12, 74] which, in Fig. 8, comprise the main column overhead accumulator, the wet gas compressors, and the debutanizer complex. From this section there is derived a light (C_2^-) gas stream (16) which may be used as refinery fuel or sent to a gas plant [71], a C_3/C_4 cut rich in propylene and butenes (17) for use as alkylation plant feed, and a C_5^+ (debutanized) gasoline cut (18), the quality of which can be continuously monitored by an on-line octane analyzer [93]. Figure 8 shows three sidecut streams produced from the main column, including a heavy naphtha (optional) (19), a light cycle (fuel) oil (20), and a heavy cycle (fuel) oil (21). The HCO can be recycled to the riser [12], i.e., via (22), but recycle ratios are lower because of the ascendency of the zeolite catalysts [67, 74, 94] (with their greater first pass conversion). Further, additional heat removal facilities have frequently been added to the main column overhead system to cope with the increased gasoline load produced with zeolites [94, 95]. The precise gasoline endpoint, and boiling range of the naphtha and light and heavy cycle oils, depend on the current product specifications, feedstock composition, etc.

Quenching of the reactor effluent (7) is effected by cooling the main column bottoms (4) (i.e., via exchange with feed, reboiling downstream towers, and generating steam) and returning it to the baffle section of the column (above the feed inlet point) [74, 96]. A portion of the main column bottoms passes to the slurry settler, a cone-bottomed vessel to which a stream of cold heavy cycle oil (23) is added; separation by gravity gives a recycleable bottoms fraction (3) in which catalyst has been concentrated, and an overhead stream, clarified slurry oil (24) [12, 74, 96]. The latter stream (retaining only minor quantities of catalyst fines) is suitable for selected fuel applications or use as carbon black feed.

D. Unsaturated Gas Plant

An unsaturated gas plant is an assembly of light hydrocarbon fractionators and absorbers for further processing of olefinic gas grom catalytic cracking, coking, etc. [71]; handling of unsaturated refinery gas streams is usually segregated from that of saturate gases. A representative unsaturated gas plant flow scheme [71] is depicted in Fig. 12. The processing of the main column overhead outlined in

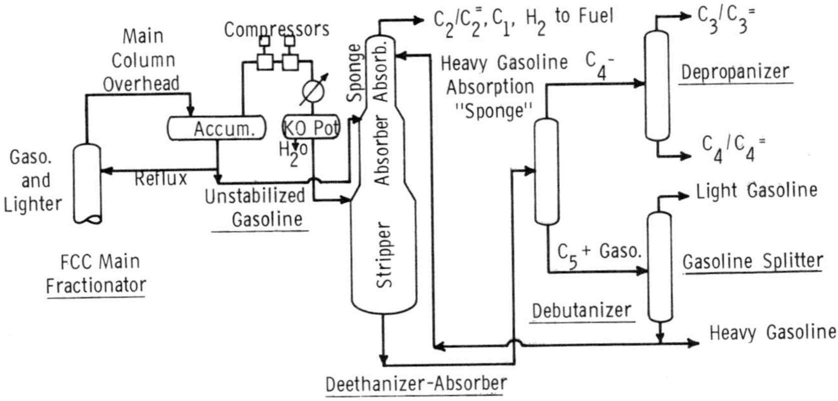

FIG. 12. Design of an unsaturated gas plant. Many variations are possible, depending on refinery product slate and economics. Heavy products can be taken off first or the reverse.

Fig. 8 (depicted within the battery limits of the FCC complex) could well be in a gas plant such as that shown in Fig. 12. The advent of the zeolite catalysts necessitated only small modifications in existing gas plants [77, 95], although some fractionation efficiency was occasionally sacrificed because of the higher yields of gasoline [77]. Further, the much lower ratio of C_2 and lighter to gasoline permits significantly lower pressure (and compression costs) in recovery of C_3 fractions [74].

E. Alkylation Unit

Paraffin alkylation, involving reaction of isobutane with butene, propylene, or mixed C_3/C_4 olefin streams (Fig. 8, 17), is usually practiced in refineries with catalytic crackers [97, 98]. High octane, highly branched, paraffinic gasoline blending stocks are synthesized by recombination of low molecular weight "degradation" products of cracking in the presence of liquid hydrofluoric acid or sulfuric acid catalysts.

It is appropriate to include a brief discussion on product distribution at this point. Table 3 gives a representative yield structure from short contact time riser FCC cracking of a fairly clean feed over a zeolite catalyst [99]. The last column in Table 3 relates the composition of this breakdown to the refinery streams in

TABLE 3

Yield Structure from Short Contact Time Riser FCC Pilot Plant Cracking of Heavy Vacuum Gas Oil over a Commercial Fluid Zeolite Catalyst[a]

Reactor inlet temperature, °F yields[b]	900° F		1000° F		Reference stream in Fig. 8
	Wt%	Vol%	Wt%	Vol%	
Conversion	77.15	79.41	78.35	80.57	(16)
Hydrogen	0.02	-	0.05	-	
Methane	0.31	-	0.71	-	
Ethylene	0.42	-	0.59	-	
Ethane	0.34	-	0.56	-	
Propylene	2.72	4.60	4.81	8.14	(17)
Propane	1.28	2.22	1.14	1.99	
Butenes	3.52	5.11	6.78	9.83	
i-Butane	5.17	8.11	3.86	6.06	
n-Butane	1.24	1.89	1.11	1.68	
Pentenes	1.47	1.99	3.57	4.83	(18)
i-Pentane	6.11	8.65	3.45	4.89	
n-Pentane	1.32	1.86	.83	1.17	
C_6-356 at 90% ASTM gasoline	48.42	56.64	48.31	55.55	
Light fuel oil[c]	15.20	13.69	15.80	14.19	(20)
Heavy fuel oil[c]	7.65	6.90	5.85	5.24	(21), (24)
Coke	4.81	-	2.58	-	(8)
	100.00		100.00		

[a]Data from Ref. 99; feed is hydrotreated heavy vacuum gas oil (28.4°API, 405 molecular weight, 13.02% H, 0.037% total nitrogen, < 0.3 ppm total metals; 4 to 10 sec catalyst residence time; 97 to 98% total recoveries.

[b]No loss basis on feed.

[c]LFO/HFO separation estimated from ASTM distillation curves of total cycle oil; LFO ~400 to 650° F, HFO ~650 to 950° F.

Fig. 8, with the understanding that there is often considerable variation in and overlap between cuts, and that there is no naphtha cut in Table 3. Alternatively, the C_3's could be allocated to LPG. Note that coke on feed is proportional to the Δ carbon (carbon on spent less that on regenerated catalyst) and the catalyst/oil ratio [12], and ultimately appears in regenerator flue gas as CO and CO_2. Commercial yields from zeolitic cracking in TCC (bead catalyst) configuration [100, 101], FCC with dense bed zone [101], and all-riser, short contact time FCC [19, 77] have been reported. The generally high olefin levels (and increase in olefinicity at higher temperature) shown in Table 3 have also been observed in commercial tests [19, 77]. Further discussion of yield structure will be developed later as it relates to the reaction variables and the chemical mechanism.

VI. CHARACTERIZATION, SOURCES, AND PREPARATION OF FCC FEEDSTOCKS

A. Characterization

It is necessary to obtain reliable characterization of a catalytic cracker feedstock in order to develop credible correlations for accurate FCC design and product yield quality predictions [64, 102, 103]. It is essential to determine certain elemental analyses and physical measurements [102, 103] such as those listed for a variety of feedstocks in Table 4. For more accurate yield predictions, however, more refined characterization of key component types—above and beyond mere knowledge of class (P, O, N, A) composition—is required [64, 103, 104, 109-111]. Of particular importance is quantitation of mono- and polynuclear aromatic distribution [103], since members of the three and four-plus ring fractions significantly affect the coke-forming tendencies of the feed [64]. Crackability, the relative cracking tendency of a feedstock at a fixed set of operating conditions [64], also decreases as the quantity of polynuclear aromatics (or aromatic/naphthene ratio) increases [67, 110, 112]. This is reflected by the parameter C_A, the percent of carbon in aromatic rings (calculated by the n-d-m method) [67, 91]. C_N and C_P, then, represent the percent of carbon in naphthenes and paraffins, respectively [91].

The aniline point is a measurement which is inversely related to the aromatic content, and is a key parameter in diesel index (D.I. = aniline point, °F X °API/100) [102]; the refractive index (n) increases with aromaticity. The K value (a function of mean average boiling point and specific gravity) [102] characterizes feeds as highly crackable (generally paraffinic, K > 12.0), intermediate (i.e., naphthenic, K 11.5 to 11.6), or refractory (generally aromatic, K < 11.3) [92, 102]. The catalyst-poisoning effect of nitrogenous bases is well known [104, 113], but no major deactivation generally occurs at nitrogen contents in the range of 0.03 to 0.09 wt% [112], nor does organic sulfur (beyond its role as a pollutant) seem to significantly affect crackability. Bromine number relates to olefin content, and viscosity values (i.e., KV at several temperatures) are required in gauging flow properties and calibrating orifice meters.

Conradson carbon residue (CCR) derives from a destructive distillation test (where the oil is burned in insufficient air) [102]. It is representative of the tendency of the feed to form coke by deposition of high boiling, refractory feed components—such as asphaltenes and resins [29] on the catalyst [28, 102]. This residue formation is probably thermal [29], and is to be distinguished from the coke formed catalytically during cracking [28, 102]. A CCR value of about 0.25 would be favorably low [102], but values in atmospheric and vacuum residua (see Table 4) may be much higher. For a number of gas oils in the low to intermediate boiling range, a certain portion of the carbon found deposited on the catalyst appears to be independent of the catalyst type or CCR, instead correlating with the basic nitrogen and average molecular weight of the feed; formation of this "additive carbon" [114] is inversely proportional to temperature.

Among the metals, nickel (and copper) are far more deleterious catalyst poisons than vanadium, and iron is the least problem of all [26, 59, 115, 116]. For metals in the feed, Nelson [117] formulates nickel equivalent (in ppm) as Ni + V/4.8 + Cu/1.23 + Fe/7.1. About 0.2 to 1.3 ppm of Ni-equivalent is representative of the level found in most FCC gas oil feeds [59].

B. Sources and Preparation

Whole crude [29] and fractions thereof from almost every boiling range [118] have been exposed to cracking conditions, with reactants ranging from light gases such as butene (co-fed with gas oil [119]) through residual bottoms [72, 120, 121]. The greatest emphasis has been on molecular weight reduction of heavy fractions. Sections III and IV serve as a frame of reference for this discussion. Because there are many alternative sources of FCC feeds, the refiner has great flexibility in their manipulation, not only in terms of being able to utilize variable or permanently different sources of supply, but also in relation to meeting seasonal or otherwise changing demand patterns.

TABLE 4

Inspections of a Variety of FCC Feedstocks

Property	A MCGO	B WTGO	C South American waxy GO	D California GO	E Gachsaran atm resid	F Gachsaran vacuum resid	G Gachsaran vacuum resid, deep deasphalted	H Cycle stock	I California coker GO
Gravity, °API (60° F)	29.6	27.2	33.6	18.8	16.05	5.9	16.0	-	29.8
Density, g/cc (60° F)	0.8783	0.8916	0.8571	0.9415	0.959	1.0298	0.9593	-	0.8772
Molecular weight	312	-		-	-	-	-	264	-
Pour point, °F	80	-	120	-	-	-	-	-	-20
Aniline no., °F	181	205	236	151	-	-	-	-	94
Conradson carbon residue, wt%	0.19	0.45	-	0.21[a]	9.4	22.2[b]	6.3[c]	0.96	-
Hydrogen, wt%	12.48	-	-	-	-	-	-	8.83	-
Sulfur, wt%	0.56	0.3	0.19	0.88	2.6	2.66	1.89	0.95	1.38
Nitrogen, wt%									
Total	0.07	0.05	0.04	0.38	-	-	-	0.47	0.25
Basic	0.03	-	-	-	-	-	-	0.04	-

Metals, ppm									
Ni	0.17		-			120	7.6	-	-
V	0.26	<3		<3	170	372	8.0	-	-
Fe	-		-			-	-	-	-
ASTM distillation, °F									
IBP									
5%	559	685	762	625	-	~1000° F+	-	713	-
50%	705	796	882	815	-		-	-	-
95%	947	865	1074	935	-		-	848	-
Type analysis, wt%									
Paraffins	28.5	-	-	-	-	-	-	4.3	-
Naphthenes	36.2	-	-	-	-	-	-	19.4	-
Aromatics	35.3	-	-	-	-	-	-	76.3	-
% C_A	14.3	-	-	-	-	-	-	-	-
K value	-	12.1	12.9	11.5	-	-	-	-	-
Refs.	18, 104, 105	106	107	106	72	108	108	104	100

[a] Ramsbottom carbon residue.
[b] 9.2 wt% heptane insoluble.
[c] <0.006 wt% heptane insoluble.

1. Feedstocks Derived from Physical Separation Processes

Straight Run Distillates

A wide range of straight run (SR) or virgin atmospheric and vacuum gas oil (distillate) fractions (see Fig. 5 for boiling range) have been characterized and their cracking behavior studied in both commercial and pilot plant units. Included are stocks such as light East Texas (LETGO, C_{11} to C_{19}) [122, 123], West Texas (WTGO) [105, 106, 124-128], Mid Continent (MCGO) [17, 18, 28, 104, 105, 126, 129], California (naphthenic-aromatic, refractory) [100, 106, 124, 130], Canadian [73, 131], South American waxy (high pour point, paraffinic, crackable) [107], Mid-East sour (2.48% sulfur) [39], Arab light vacuum [47], Gachsaran [100], and other [109, 112] gas oils. Inspections of such gas oils are shown in Table 4 A-D. Even light straight run paraffinic naphtha has been cracked in an FCC system [132, 133], with the objective of increasing aromatic content and octane number rather than molecular weight reduction.

Atmospheric Residua and Related Fractions

In most existing refinery configurations (i.e., Fig. 7), atmospheric tower bottoms (otherwise known as residua, reduced crude, or topped crude) are sent to a vacuum tower. Here, metallic contaminants and high molecular weight asphaltenic and heteroatom impurities are concentrated in the vacuum residua [39, 134], and the FCC catalyst is exposed to a relatively clean feed (<3 to 4 ppm [50] and preferably <1 ppm [26] of Ni + V). In other cases, direct cracking of atmospheric resid has been reported [39, 50, 72, 120, 121]. This approach poses a more formidable challenge in feed properties [36, 59, 72, 135], since typical resids may contain high metals poison levels (170 ppm in the Gachsaran resid [72] shown in Table 4 E), large amounts of asphaltene-type coke precursors (to some extent reflected in from 5 to 20% CCR levels [120, 135], and sulfur levels as high as 2.7% [39, 120]. The cracking of high metals (Ni, 16 ppm; V, 44 ppm) whole Iranian crude reported by Masologites and Beckberger [29] can be appropriately mentioned here, since this feed has 3.6% CCR and contains 19% 950° F+, 9.5 °API residuum (and 25.5% 650 to 950° F heavy gas oil).

Solvent Deasphalting

Thus direct cracking of resid—with the resultant necessity of coping with high catalyst metals and coke levels [39, 72, 134]—is one approach to utilize "the bottom of the barrel." Another approach, one that takes advantage of the insolubility of asphaltenes in propane and light paraffins [34-36], is solvent deasphalting [29, 36, 39, 57, 108, 118, 134]. Asphaltenes, some resins, and associated organometallic components are concentrated into a bottoms "asphalt" fraction [29, 136, 137], producing a deasphalted extract oil with significantly reduced metal and coke precursor concentration. Table 4 F and G show inspections of a Gachsaran vacuum resid before and after a deep-deasphalting operation [108]. The lower viscosity extracts may then be cracked directly [39, 134, 136] or hydrotreated and then cracked [57]. Slops from lube processing may also be cracked.

2. Feedstocks from Physicochemical Preprocessing

Recycle Oils from Catalytic Cracking and Coker Stocks

As described in Section V-A, the fresh feed to the riser may be mixed with slurry settler bottoms [96] and/or heavy cycle oil stocks [94, 124, 138], and the combined feed subjected to cracking. The combined feed ratio (CFR) is defined as the (weight) ratio of fresh feed plus recycle streams to fresh feed [94]. The effect of CFR, i.e., the amount of recycle, on yield structure for a number of possible refinery situations has been described [124, 138]. Because the more active zeolite catalysts afforded higher first pass conversion of fresh feed, a significant drop in recycle requirements and cycle oil/bottoms yields occurred, with average CFR dropping from about 1.41 in 1964 to 1.27 in 1972 [94]. Cycle stocks are much denser, more aromatic, and resistant to further cracking than fresh feeds [67, 103, 124, 126, 130] because the more readily cracked components—including paraffins, naphthenes, and aromatic side chains—have already been removed in the first pass operation [104, 109, 110]. Many cycle stocks contain as much as 2 to 5 wt% sulfur [39]. Properties of a typical FCC cycle stock (HCO) reported by Voltz et al. [104] are given in Table 4 H. Use of clarified slurry oil as FCC feedstock leads to large amounts of coke formation [94].

Coker gas oils, which still represent a substantial portion of the feed to an FCC [94, 103, 126], are also generally more unreactive than virgin gas oils [94, 109, 110]. Reflecting their high severity thermal cracking origins, coker stocks often contain appreciable quantities of olefins [110], and they are commonly high in aromatics and sulfur [126]. It is also more difficult to predict the reactivity of coker and recycle stocks [104]. Some properties of a coker gas oil are shown in Table 4 I. Slops from chemical plants, i.e., heavy pyrolysis oils, and heavy oils from the visbreaker [139] may also be routed to an FCC. Also, oils derived from pyrolysis of coal and shale oil from thermal shale retorting have been catalytically cracked

[140]; in addition to the sulfur problem, these have more oxygen and nitrogen than petroleum stocks. Although previously thermally cracked, coker stocks are classified as fresh feed [141].

Hydrotreated Stocks

A variety of potential FCC feedstocks has been upgraded by catalytic hydrotreating; these include straight run gas oils [39, 64, 130], atmospheric residua [26, 59], vacuum gas oils [47, 59, 64, 67, 142-144], vacuum resid [39], solvent-deasphalted oils [59, 64, 143], coker stocks [143, 144], Utah coal-derived oils [145] (with their low hydrogen content and high concentration of aromatics and heteroatoms), and light [67] and heavy FCC recycle stocks [50, 59, 64, 130, 142, 144, 146]. Cycle stocks, particularly from high-conversion zeolite FCC operations, are highly aromatic and require severe [94] hydrogenation to saturate polynuclear aromatics to mononuclear [59] prior to recycle. This appreciably improves their crackability.

The changes in feedstock properties obtained upon severe hydrotreating of a 650° F+ Kuwait atmospheric residua [59] are listed in Table 5. Sulfur removal in hydrotreating is generally much easier than nitrogen removal [142], the severity for denitrogenation being very much a function of the feed boiling range, type of nitrogen heterocyclic, and feed source [50, 142, 147]. Note in Table 5 that at conditions where 97% desulfurization and 99% demetallization (removal of trace metals, i.e., Ni-, V-organometallics [59, 130]) occurred, only 50% of the nitrogen was removed. Also, with mild hydrotreating, there is sometimes an increase in basic nitrogen, with consequent poisoning of cracking catalyst [47-49].

Hydrotreating generally improves crackability [39, 47, 59, 130], with the extent of improvement usually proportional to the amount of hydrogen consumption [39, 142, 147]; and the cracking patterns are often similar to those of distillates [59, 99]. The hydrotreating (HDT)-FCC combination [39, 57, 59, 118, 126, 142, 148, 149], with HDT considered as feed preparation or as a pollutant cleanup process [50], is generally associated with improved selectivity and product quality in cracking [47]. Thus higher conversion and gasoline yield, and lower selectivity to coke have commonly been reported in FCC cracking of hydrogenated stocks [39, 47, 59, 130, 142, 147, 150, 151]; also, more favorable light gas distribution, including higher isobutane yields, has been observed [39, 147]. Improved quality of cracked products, ranging from gas through coke, notably in lower sulfur and nitrogen contents [47, 59, 130, 152], allows meeting ultimate SO_2 and NO_x pollution specifications [59, 92, 118, 130]; improved gasoline lead susceptibility has also been observed [118, 130]. A comparison of the HDT-FCC [153] combination with hydrocracking has

TABLE 5

Effect of Severe Hydrotreating on Properties of Kuwait Atmospheric Resid [59]

Property[a]	Feed, Kuwait 650° F+ atm resid	Hydrotreated products 460°+ (topped) FCC charge
Sulfur, wt%	3.8	0.1
Nitrogen, wt%	0.21	0.11
Ni, ppm	15	0.2
V, ppm	45	0.1
Ni-equivalent (Ni + 1/5 V)	24	0.22
Gravity, °API (60° F)	16.6	26
Density, g/cc	0.9554	0.8984
Carbon residua, wt%[b]	8.3	2.2
Chemical H_2-consumption, scf/bbl	-	980

[a] Aniline point increases, pour point and viscosity decrease.
[b] Ramsbottom.

been reported [146]. Milstein et al. [153] also note that coupling of HDT and FCC could help reduce United States dependence on foreign crude.

More specifically, hydrotreating converts asphaltenes and potential coke-forming material [26, 59] and saturates polynuclear aromatic ring systems [50, 130, 142] (preferably to mononuclear aromatics [47, 67] and more crackable naphthenes and paraffins [130]) so that less coke is formed upon cracking [130]. Much of the hydrogen consumption in HDT can be related to saturation of polynuclear aromatics [130, 147] (and sulfur and nitrogen heterocyclics [130, 142, 147]). Polynuclear saturation may require more severity than hydrodesulfurization or demetallization [142]. A detailed study of the saturation patterns of polynuclear aromatics (di- through penta-) in a California vacuum gas oil/cycle oil mix was reported by McCulloch

[147]. A number of studies emphasizing the effect of HDT severity on FCC process variables, crackability, yields/selectivity, and product quality have been reported [99, 130, 142, 147, 152]. Hydrotreating catalysts are designed to accept large amounts of deposited metals (i.e., act as a metals sink), thereby preventing contamination of the FCC catalyst, the performance of which is much more sensitive to metals [142].

VII. PROPERTIES AND CHARACTERIZATION OF FCC CATALYSTS

A. Background and Scope

As noted in Table 1, synthetic silica-alumina was introduced as a cracking catalyst about 1940 and, in one modification or another, dominated the cracking catalyst field until the early 1960s and the introduction of the zeolites [13, 17, 18]. However, over the decade of the 1950s, and particularly during the latter half, only incremental improvements in selectivity and regenerability of the amorphous catalysts (e.g., via silica-magnesia variations [128, 154, 155]) were achieved, and it remained for the zeolite catalysts to truly revolutionize the cracking catalyst field.

Plank and Rosinski (see the series of patents issued to Mobil Oil [156-161]) discovered that incorporation of a minor proportion of zeolite (such as rare-earth cation exchanged X or Y synthetic faujasite) in a matrix such as silica-alumina, silica, or silica-clay, produced compositions of matter that, over more than a decade, have met, to a large extent, the most severe requirements [13, 17, 18] for successful cracking catalysts, including:

1. High activity (allowing economic throughput rates)
2. Reasonable activity stability (resistance to time-dependent deactivation by coke)
3. High selectivity to gasoline vs coke and dry gas
4. Thermal and hydrothermal stability (ruggedness to withstand the regenerator environment)
5. Accessibility (sufficient diffusivity and internal porosity to allow small, economic reactor volumes)
6. Attrition resistance
7. Resistance to poisons (metals, nitrogen, etc.)
8. Acceptably low cost

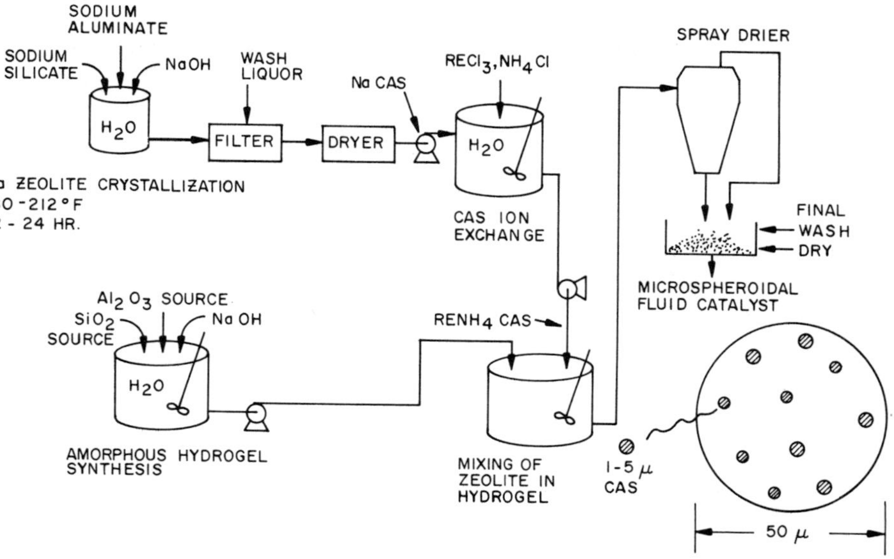

FIG. 13. Simplified conceptual scheme for manufacture of commercial fluid cracking catalyst (XZ-25). Modified after flow sheets in Koffler [181].

In this section we attempt to focus on highlights of the zeolite cracking catalysts, limiting our discussion of the amorphous silica-alumina type to selected comparisons only, as these have been well described by Sittig [10], Voge [11], and others [45, 122, 128, 162-180].

B. Manufacture of FCC Catalysts

As visualized in Fig. 13, the major operations in the manufacture of a commercial zeolite cracking catalyst [182] may include synthesis and ion exchange of the zeolite component, preparation of the amorphous gel, intimate admixture of the zeolite in the gel matrix, and spray drying.

Manufacture of zeolites commercially involves combination of readily available, inexpensive, raw materials, including a silica source, an alumina source, and a strong base such as sodium hydroxide [13, 181, 183, 184]. These materials, in proper stoichiometric ratios, molarity, and state (i.e., solution or colloidal dispersion), are heated at a given temperature for a time period specified to give maximum purity of the desired structure. Approximate conditions for the preparation of synthetic faujasite X are shown in Fig. 13 [15, 16, 181]. As initially crystallized, most zeolites contain sodium cations and are in the form of finely divided particles, on the order of 1 to 7 μm [13, 101, 156, 185]. Excellent overviews of the complex area of zeolite synthesis have been published by Breck [183] and Flanigen [184].

In the next phase the preponderance of sodium may be removed by cation exchange with a medium containing rare-earth cations, or rare-earth cations plus hydrogen ion, or a cation such as ammonium ion capable of conversion to hydrogen ion [13, 17, 18, 127, 156-161]. Inexpensive rare-earth sources (as monazite or bastnasite sands), generally as the chloride salts, are utilized [13, 156-161, 181]; these are rich in cerium, lanthanum, neodymium, and praseodymium, with lesser amounts of samarium, gadolinium, etc. [181]. Comprehensive reviews on zeolitic ion exchange have been published by Sherry [186, 187].

Finally, a siliceous inorganic oxide-type sol or gel (ultimately the matrix) is produced by methods well known in the art [188] (concisely described in a review by Oblad [13]), into which the micron-sized zeolite powder is uniformly dispersed [156, 161], producing a mechanically stable, rugged, microspheroidal composite fluid catalyst after spray drying. Current commercial manufacturing practices vary widely, of course [13].

C. Some Key Structural Features of Zeolites

The structure of zeolites (particularly rare-earth exchanged synthetic faujasites) has been discussed intensively in the literature in relation to their role in catalytic cracking [13, 17, 18, 62, 65, 76, 101, 106, 127, 131, 181, 189, 190]. Particularly relevant are overviews by Ciapetta [191], Meier and Olson [192], Oblad [13], Rabo and Poutsma [193], Smith [194], Turkevich and Ono [195], Venuto and Landis [196], and Weisz [197]. Salient features are summarized below.

Zeolites may be viewed as crystalline, inorganic polymers comprising an anionic lattice of AlO_4 and SiO_4 tetrahedra linked by shared oxygen (Fig. 14). A cubo-octahedron or sodalite cage (Fig. 15A) (24 Al, Si ions, 36 oxygen anions, 8 hexagonal, and 6 square faces) represents the simplest structural building block. When sodalite units are joined in tetrahedral mode (Fig. 16B) (by bridging oxygens across the hexagonal faces), a rigid, three-dimensional reticulum—the faujasite structure (Fig. 16A)—is generated, which upon dehydration

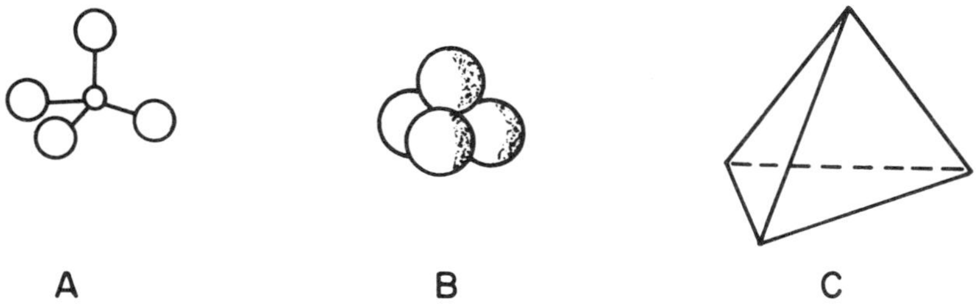

FIG. 14. Elementary representation of TO_4 (AlO_4/SiO_4) groupings in zeolites. (A) Skeletal tetrahedron. (B) Packed spheres, the "sea of oxygen model." (C) Solid tetrahedron; units are joined by sharing apical oxygen atoms.

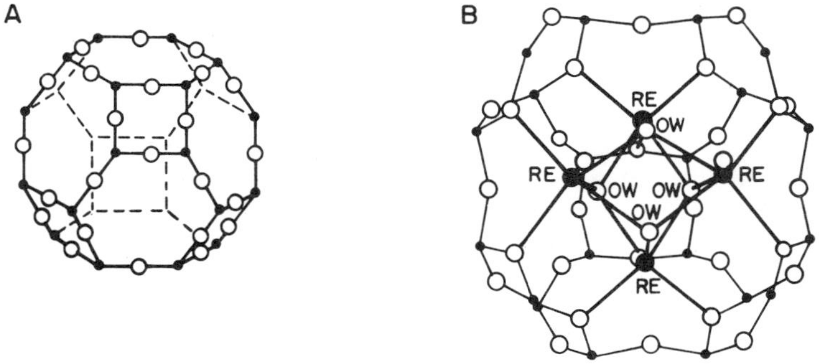

FIG. 15. (A) Geometric representation of sodalite cage as a truncated octahedron. (B) Sodalite cage complex containing four rare earth cations. After Olson, Kokotailo, and Charnell [198]. (•) = Si, Al; (o) = oxygen; (●) = rare earth.

affords a continuous channel system of uniform pores. Very high internal surface area (~500 to 800 m^2/g) becomes available. Thus for a catalytic reaction (like cracking) with an activation energy of 30 kcal/mole or higher to have commercially feasible throughput, the guidepost suggested by Weisz [197]—"the availability of a surface of at least 1 to several hundreds of m^2 per cm^3 of catalyst volume"— can be met.

One consequence of the atomic configuration in faujasite is the existence of a periodic pattern of tetrahedrally oriented large- and small-pore systems (Fig. 16A, 16B). The former comprises the

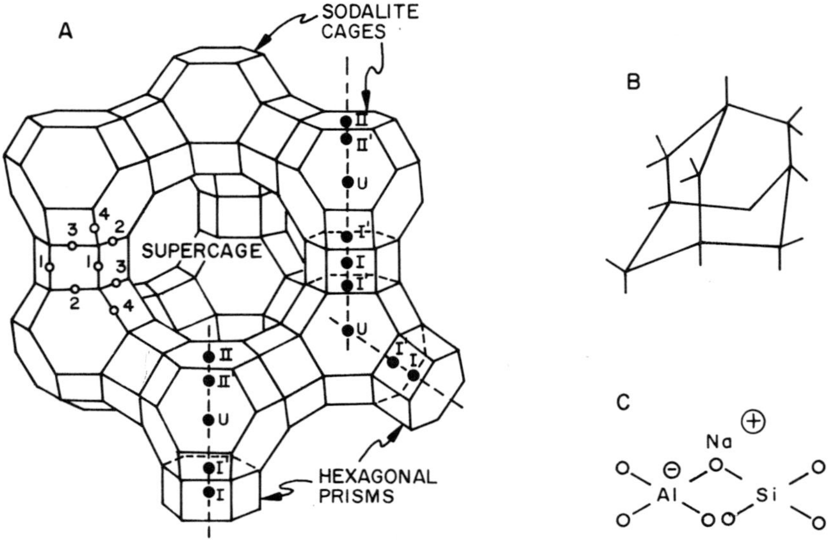

FIG. 16. (A) Schematic of faujasite framework showing oxygen type (O) and nonframework (●) locations; after Scherzer, Bass, and Hunter [199]. (B) Adamantanoid pore structure in faujasite. (C) Basic unit of cation exchange/functionality.

spherical supercages (which access 4 identical cavities via 8 to 9 Å diameter 12-ring windows) and the latter the internal voids of the sodalite cages and hexagonal prisms. Facile sorption (rapid kinetics and rectangular isotherms) of potential reactant molecules with critical dimensions less than those of the pore diameters is generally observed [196]; these can include some C_{18} to C_{25} mono-, di-, and trinuclear aromatics present in heavy gas oil [200], although exclusion by size/shape will of course occur with very large/bulky molecules [99, 196, 201, 202] (another extension of the molecular size-selectivity principle [197, 203, 204]). The small pore system (entry via the ~2.5 Å 6-ring of sodalite cage) is generally inaccessible to organic molecules, but permits sorption of water or ammonia, and can be involved with an intricate variety of ion sievings [186, 187]. A plurality of crystallographically-distinct oxygen types is present (1 to 4 in Fig. 16A), and a complex hierarchy of cation siting has been elucidated [194]; sodium cations in dehydrated Y-type faujasite are usually located at the four major positions, Sites I, I', II, and II'

[205] (Fig. 16A), and to maintain electrical neutrality, are associated with the unit negative charge of an AlO_4^- cluster (Fig. 16C). U marks the center of the truncated octahedron.

In his early work on rare-earth ion exchange, Sherry [206] noted that La^{3+} ions could not easily replace 16 Na^+ ions/unit cell of zeolites X and Y at 25° C, reflecting the ability of the hydrated La^{3+} ion (d = 7.92 Å) to exchange with supercage sodium ions, and its inability to penetrate the 6-ring portal (d = 2.5 Å) of the sodalite cage [187]. Upon stripping the water of hydration (as with ion exchange at higher temperatures, calcination, or vacuum dehydration), the "bare" La^{3+} cations (d = 2.3 Å) could then pass through the window between the large and small cages [187, 198]. Olson, Kokotailo, and Charnell [198] showed that each rare-earth cation could coordinate with 3 framework and as many as 3 water oxygens in approximately octahedral coordination to form an extremely stable complex of high cation density within the sodalite cages (Fig. 15B), leaving the supercage void of metal cations. Similar observations were made by J. V. Smith and associates [194]. In a later x-ray analysis of dehydrated CeX, Hunter and Scherzer [207] showed that the zeolite lattice was distinctly altered upon oxidation of Ce^{3+} to Ce^{4+}; Scherzer, Bass, and Hunter, in a study of hydrothermally treated LaY [199], demonstrated that steaming causes partial framework dealumination and possible structural rearrangement.

The nature of the acidity in zeolites has been the subject of intense investigation and speculation (see for instance the work of Rabo and Poutsma [193], Turkevich and Ono [195], Venuto and Landis [196], Ward [208], Beaumont, Barthomeuf, and Trambouze [209], Moscou [210], Hall [211], and their associates). A wide variety of techniques has been employed in the characterization of these acidic properties, including IR spectroscopy (before and after adsorption of basic species), titration with diverse bases (such as amines or $LiAlH_4$), the study of acid site strength and distribution by means of differential thermal or thermogravimetric analysis, calorimetric or chromatographic estimation of heats of adsorption, the use of Hammett-type chromophoric indicators, electrophilic hydrogen-deuterium exchange, metal-catalyzed "titration" with H_2 or D_2, and the use of NMR methods.

Acidity has been commonly ascribed to Brönsted, Lewis, or "defect" sites associated intrinsically with the crystalline lattice, cation fields, or even sites generated from chemisorbed polymeric "residue." Such sites have been postulated as deriving from severe dehydration of the crystal, thermal decomposition of ammonium of alkyl-ammonium cations, hydrolysis of multivalent cations, exchange with protonic acids, dehydroxylation of the lattice, and numerous other pathways. Regardless of the pathway of origin, however, most of the carbonium-ion type catalytic activity exhibited by zeolites appears to derive from strongly acidic protons [196, 205, 210, 211]. But, is there a satisfactory answer to the question of why zeolite catalysis should be different from that with other solid Brönsted acids? Barrer, Bultitude, and Sutherland [212], at an early date, emphasized that no part of any zeolitic channel was wide enough for occluded substrates to escape the (electrostatic) field potentials derived from the surrounding lattice components. Rabo and Kasai [205], in addressing this question, noted that the voids created by lattice dehydration are large (on the atomic scale) and hence less desirable in terms of an ionic lattice. They suggest that there should be a driving force to fill the intracrystalline void with polar/ionic species—with induced positive and negative charges geometrically arranged according to the preference of the whole crystal—and that protonic acidity is enhanced by the overall gain of crystal (Madelung) energy. Viewed in quite a different perspective [248], this enhancement may reflect a subtle "band structure" arising from the strategic arrangement of atoms in the highly ordered, rigid, zeolitic network.

D. Interactions between Cracking Catalyst Structure and Process Requirements

1. The Activity Factor

The activity advantage of zeolite cracking catalysts over the amorphous silica-alumina types has been well documented [100, 105, 127, 131, 190]; in early gas oil cracking studies, Plank, Rosinski, and Hawthorne [18] noted that rare-earth-exchanged X-type faujasite was more than 100 times as active as silica-alumina. In fact, Weisz and associates [168, 197, 213], using the α-test (n-hexane cracking, microreactor), demonstrated zeolitic activities 10^3 to 10^4 times greater than those of conventional catalysts.

The activity of modern zeolite cracking catalysts has improved to the point where, for the most part, only 1 to 4 sec of contact time are required to effect substantially complete conversion of the nonaromatic portion of the feed [65]. With the earlier TCC bead catalysts, the availability of this high activity actually engendered the risk of moving into an undesirable diffusion regime [168], and provided incentive for development of fluidized systems. As a convenient rule of thumb, Weisz [197] estimates that a chemical transformation rate of $\sim 10^{-6}$ mole/sec/cm^3 of catalyst volume is required for an (economically) feasible commercial throughput (rough, average conversion rates on the order of 10^{-4} to 10^{-5} mole/sec/cm^3 can be calculated for the equilibrium zeolite catalyst cracking data [99] described earlier in Table 3).

Note that low sodium content is a requisite for high activity [101] (the sodium forms do not have the necessary catalytic activity for effective hydrocarbon conversion [191]), and that rare-earth Y-type faujasites show higher activity than the X-type [76]. That conventional silica-alumina catalysts are poisoned by sodium ions is well known [214]. Also, unsteamed catalysts have exceptionally high activity, and steaming—prior to use or while in FCC circulation—gives substantial reduction of activity. Detailed studies have been made on the effect of (rare-earth) zeolite concentration in the matrix on activity, selectivity, etc. [106, 125]. Finally, Plank, Rosinski, and Hawthorne [18] have demonstrated that rare-earth X crystalline aluminosilicate continued to function effectively at coke levels which would completely deactivate conventional catalysts. Sophisticated kinetic analyses of the time-dependent decay of catalyst activity (fluidized dense bed reactor) have been reported by Nace, Voltz, and Weekman [109, 110].

2. Selectivity and Product Quality

The outstanding selectivity characteristics of zeolites were recognized at an early stage [18, 100, 101, 105, 107, 127, 131, 190].

The yields in Table 6, from the pioneering work of Plank and Rosinski [17], are representative, showing the vastly superior selectivity of zeolites over silica-alumina, namely production of large amounts of incremental gasoline at the expense of C_4's, dry gas, and coke. The enhanced selectivity is particularly prominent in crystalline aluminosilicates exchanged to low sodium levels with both rare-earth cations and proton precursors [156-161]. A large increase in commercial cracker throughput without the need for large capital expenditure was thus enabled [181]. The increased ratio of gasoline/gas + coke reflects the direction of cleavage of higher hydrocarbons to the C_5 to C_{12} range by faujasite [17, 18].

A detailed differential component breakdown over commercial fluid catalysts from the work of Blazek [107] is given in Table 7. Data are for cracking of a waxy gas oil (see Table 4 C above for inspections), the type of paraffinic feedstock on which the advantages of zeolites are most evident. General product trends are typical: note the ~20% reduction in coke with the zeolite, a great benefit in a coke-limited unit; H_2 and C_1 and C_2's are reduced by about 40%; C_3 and C_4 olefins are greatly reduced, reflecting the superior hydrogen transfer facility of zeolites, while C_3 and C_4 saturates show

TABLE 6

Selectivity Advantage of Rare-Earth Zeolite Catalysts over Amorphous Catalyst at Same Conversion Level

	Δ advantage over silica-alumina[a]	
	REHX[b]	10% REHX in matrix[c]
C_5^+ gasoline, vol%	+7.1	+5.9
C_4's, vol%	-3.5	-1.3
Dry gas, wt%	-2.1	-1.5
Coke, wt%	-2.1	-2.2

[a]Fixed bed cracking, mid-continent gas oil; data from Ref. 17.
[b]Na (0.22%), $(RE)_2O_3$ (26.9%), Al_2O_3 (29.6%), SiO_2 (41.4%).
[c]SiO_2-Al_2O_3 gel.

a slight decline (and isobutane/butane ratio increases); yields of C_5^+ gasoline and light fuel increase by 6.5 and 1.9 vol%, respectively; heavy fuel oil, however, declines by 1.9 vol% and is denser.

As shown in Fig. 17 for three very different gas oil feedstocks, gasolines produced by zeolite cracking generally have more paraffins and aromatics and less olefins [143] than those from silica-alumina [100, 190]; but whether or not they show higher octane numbers depends upon which specific members of the classes P, O, N, A are present and in what ratio; this in turn is a function of feedstock composition and process variables. Zeolitic gasolines commonly are lower in sulfur and nitrogen content and have better (storage) stability [100] and lead susceptibility [107, 143].

3. Thermal and Hydrothermal Stability

The accessibility, activity, and selectivity of a cracking catalyst must be maintained throughout countless cycles of high severity regeneration. The rate of deactivation (loss of crystallinity) is affected most critically by temperature and water (steam) partial pressure

TABLE 7

Comparison of Yield Structure for Fluid Catalytic Cracking of Waxy Gas Oil over Commercial Equilibrium Zeolite and Amorphous Catalysts[a]

Yields, at 80 vol% conversion[b]	Amorphous, high alumina	Zeolite, XZ-25	Δ change from amorphous
Hydrogen, wt%	0.08	0.04	-0.04
$C_1 + C_2$'s, wt%	3.8	2.1	-1.7
Propylene, vol%	16.1	11.8	-4.3
Propane, vol%	1.5	1.3	-0.2
Total C_3's	17.6	13.1	-4.5
Butenes, vol%	12.2	7.8	-4.4
i-Butane, vol%	7.9	7.2	-0.7
n-Butane, vol%	0.7	0.4	-0.3
Total C_4's	20.8	15.4	-5.4
C_5-390 at 90% ASTM gasoline, vol%	55.5	62.0	+6.5
Light fuel oil, vol%	4.2	6.1	+1.9
Heavy fuel oil, vol%[c]	15.8	13.9	-1.9
Coke, wt%	5.6	4.1	-1.5
Gasoline octane no., R+O	94	89.8	-4.2

[a]Data from Blazek [107]; run at 950° F and 1 atm in a fixed fluid bed.
[b]No loss basis on feed.
[c]HFO = 100 - conversion - LFO.

[101]. The faujasite-type catalysts are much more resistant to heat and steam than conventional silica-alumina [17, 18, 131], with the higher SiO_2/Al_2O_3 ratio Y-type showing even more stability than the X-type [101]. Even with the zeolites, however, the sodium concentration should be decreased to as low a residual level as possible, with trivalent rare-earth cation exchange imparting the greatest

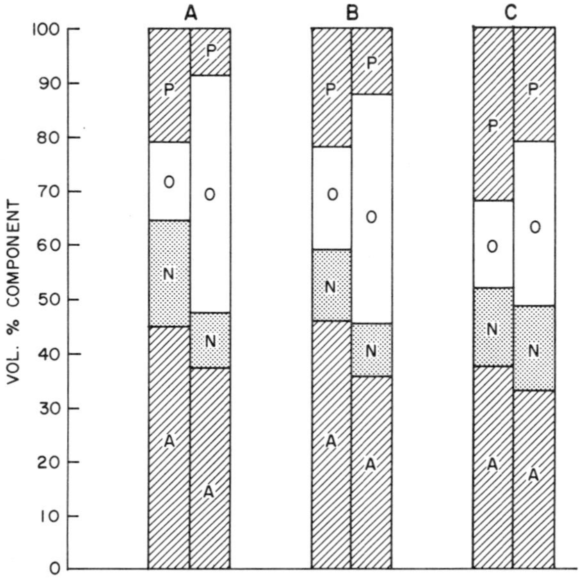

FIG. 17. Comparison of paraffin, olefin, naphthene, and aromatic (P, O, N, A) contents in C_5-375/400° F A.S.T.M. 90% point TCC gasoline from catalytic cracking of (A) California virgin gas oil, (B) California coker gas oil, (C) Gachsaran virgin gas oil. Left and right members of bar graph pairs reflect zeolitic and amorphous cracking, respectively. Data are from Ref. 100.

stability to high temperature and moisture [17, 18, 101, 181]. "Ultrastable" faujasite compositions, thermally stable with retention of high (800 to 1000 m^2/g) surface area at temperatures greater than 1000° C, have been reported [106, 191, 215]. It will be recalled that under certain circumstances the temperature within the interior of a catalyst particle may greatly exceed that at the surface as recorded in the contacting fluid phase [216-219]; thus unsuspected catalyst damage can sometimes occur during the exothermic carbon burnoff.

4. The Question of Accessibility

It is critically important that the integrity of the zeolite (and matrix) pore structure be maintained, and that deposits of carbonaceous matter, metals, or other inorganics that block access of molecules approaching and leaving zeolitic sites be minimized [65, 191]. While the question of diffusivity must not be neglected for fluid catalysts (with average particle size of 50 to 60 μm), it is of considerably greater concern with 0.3 to 0.4 cm diameter TCC beads which may have a 40 to 50 times greater diffusion path. Here, diffusivities of 15 to 30 X $10^{-3}cm^2$/sec have been reported [105, 131, 167], such levels being required [105] to take advantage of the high activity/selectivity of crystalline aluminosilicates.

With bead catalysts, not only must mass transfer aspects during the cracking reaction be considered, but diffusion inhibition during regeneration must be avoided [167].

5. Attrition Resistance

It is highly desirable to have fluid catalysts that are resistant to mechanical attrition—the formation of "fines," particles less than 20 μm [65]. With bead catalysts, a study by Weisz [220] showed that the rate of attrition was indirectly dependent upon regeneration kinetics—and diffusion/mass transport—in that differential shrinkage between the coked, core portion and the clean, shell portion of the bead would create internal mechanical strain, with increased tendencies toward attritional wear.

6. Resistance to Metals Poisoning

The tendency of metals to induce nonselective cracking in amorphous catalysts has been well defined [221-223]. Zeolite cracking catalysts have excellent resistance to contamination by vanadium, nickel, and other heavy metals [92, 131], and indeed appear to be superior to amorphous catalysts in this respect [17, 18, 28, 29, 101]. This is evidenced by the lower carbon producing factors (CPF) at a given metals loading for zeolites [29], by the fact that up to 3000 ppm effective metals the zeolites produce only one-third as much additional coke as the amorphous [29], and by the (statistical) observation that metals levels in commercial equilibrium catalysts have been allowed to increase since the ascendancy of the zeolites [101]. The faujasite catalysts are also more resistant to poisoning by organic nitrogen compounds [17, 18]. These favorable attributes of zeolites are particularly valuable with the inevitable trend toward processing of increasingly heavy feedstocks [65].

Metals may mask differences between cracking catalysts unless proper correction is made [28]. As observed by Masologites et al. [29], gross metals level cannot be used as an index of contamination, since not all deposited metals are equally effective in producing coke and hydrogen [12]. On a weight basis, nickel shows four times the coke- and hydrogen-forming tendency of vanadium [29, 224], and similar relative responses for both metals on amorphous silica-alumina have been reported [221]. Cimbalo, Foster, and Wachtel [28] define "effective metals" (in the contaminant sense) as (4 [ppm Ni] + [ppm V]) X "fraction effectiveness" which, in turn, is a function of feed type and throughput, catalyst type, age and history, and unit inventory and makeup rate [28, 29]. As an example, for systems where daily makeup rates ranges from 0.5 to 3.0% of inventory, the "fraction effectiveness" varied from about 0.21 to 0.32 [28]. Iron (which is commonly scale or "tramp" metal) and most other metals are not considered significant catalyst contaminants.

Freshly deposited metals are more active than "older" metals that have been subjected to numerous cycles in the FCC regenerator/reactor circuit [102]. Upon exposure to such repeated cycles of oxidation/reduction, the poisoning effects of metals contaminants are diminished, but those on zeolites lose their effectiveness more slowly than those on amorphous catalysts [28]. Grane, Conner, and Masologites [221], in a study of metals-poisoned silica-alumina, suggested that the repeated shift from oxidizing to reducing environment in going from regenerator to reactor may influence the oxidation state of the metals or cause sintering (and loss of surface area) with consequent deactivation. Morozov et al. [225] observe that, for a given feedstock, a greater metals buildup may occur on fluid catalysts than on bead types, because with the latter there is removal of significant amounts of surface metals by attrition. Montgomery [102] observes that sulfur in FCC feed may tend to sulfide—and thereby activate—the metals on a cracking catalyst. On the other hand, it has been observed that zeolite catalysts have excellent metals resistance, particularly in short contact time operation [92, 226]; this may occur because metal oxides on the circulating catalyst are not retained in the reaction zone long enough to become sufficiently sulfided to show strong dehydrogenation activity [92, 226].

7. The Role of the Matrix

The role of the matrix in composite catalysts has been discussed in some detail by several authors [13, 101, 125]. Matrices encompass a broad range of SiO_2-Al_2O_3 gels, clays, Al_2O_3, and other siliceous materials, and may be expected to vary in different catalysts [227]. Natural, chemically treated, and calcined clay have also been specified [158] as possible constituents. Although most fundamental studies have been conducted on zeolites alone, Oblad [13] points out that a fair amount of catalyst mechanistic patterning may also be true for matrix-zeolite blends.

In a sense, the matrix protects the small (~1 to 5 μm), relatively fragile zeolite crystallites, building the composite up to a ~50 μm microsphere and providing shape and bulk for fluidization [76] and minimizing mechanical attrition (see Fig. 13). Eastwood, Plank, and Weisz [101] observe that the high surface area matrix surrounding and diluting the zeolite crystals acts as a heat carrier or "sink," facilitating heat transfer during regeneration and cracking. The stability of the composite to high severity steaming is also enhanced [13, 101]. Some matrices may act as "sodium sinks," allowing solid-solid sodium ion exchange, thus improving stability of the zeolite particles [13, 101], i.e.,

$$\text{RE NaY} + \text{H-Gel} \rightarrow \text{RE HY} + \text{Na-Gel}$$

$$\text{RE NaY} + \text{Al-Gel} \rightarrow \text{RE AlY} + \text{Na-Gel}$$

Although it is true that matrices have catalytic properties of their own [13], the much greater zeolite activity permits a relatively small amount of molecular sieve to dominate the catalytic behavior of the composite [13, 158]. In fact, a synergistic effect has been described [13, 100] wherein enhancement of zeolite activity occurs (in that the same conversion can be obtained for a smaller amount of zeolite when incorporated in the matrix). The requirement for a matrix to possess—and retain— adequate porosity to maintain full accessibility to feed molecules and regeneration gases is critical [13, 76, 127, 185].

E. Performance Characteristics and Physiochemical Characteristics of FCC Catalysts

1. Some Aspects of Fluidization Dynamics

It is because solid fluid catalyst particles have the capacity to exhibit liquid-like flow properties when aerated that makes possible the high catalyst circulation rates utilized in FCC units. In fact, if the degree of slippage between the gases and solids is sufficient, the system can very closely parallel the dynamics of a true fluid system [10, 92], even to the extent of permitting flow across a (slide) valve. When a gas flows upward through a bed of solid particles at low velocity, it flows through the interstices without moving the bed. Such a situation is called a fixed bed. If the velocity (and pressure drop) are increased, a point will be reached where the pressure drop equals the weight of the bed and the gas will begin to lift the solids. The particles will move apart and the bed will expand in volume, becoming more loosely packed. As the gas velocity is further increased, the pressure drop remains essentially constant (equal to the weight of the bed) because the interstices open up to decrease the resistance to flow. Finally, a point is reached where the particles no longer touch. Gas bubbles begin to form, and the particles begin to move relative to each other in a turbulent manner. For FCC catalyst, the points of incipient bed lifting and incipient bubble formation occur at gas velocities of about 0.01 and 0.1 ft/sec, respectively [80]. For gas velocities at or higher than the point of incipient bubbling, the solid particles are described as a fluidized dense bed.

Further increase in the gas velocity results in larger bubbles. If the bed is narrow, the bubbles can become as wide as the vessel, and alternate rising layers of gas and solids will form. This is called slugging, and as the slugs of solids rise, individual particles fall through the gas layer and the solid slug eventually disintegrates. At very high velocities the terminal velocity of solid particles is exceeded and the particles are carried along with the gas. Such a regime is called a dilute phase with pneumatic transport of solids [228]. The various states of an aerated bed are illustrated in Fig. 18.

Whenever a fluidized bed is used, some of the particles—particularly the smaller ones (fines)—are entrained and carried out with the gas. To overcome this difficulty, cyclonic collectors (somewhat analogous to centrifuges) are employed. The gas/solids mixture enters the cylinder tangentially, gas is removed overhead from the middle, and the solids are displaced toward the wall and funneled downward toward diplegs (see Figs. 9 and 10). Key cyclone performance variables include inlet velocity, component geometry, particle size/density, and gas viscosity [63, 82-84, 88].

In a fluidized dense bed there are two ranges of gas holdup times: One for the gas that goes through the bed in bubbles and another, much longer one, for the gas that proceeds through the interstices between the particles. The gas in the bubbles tends to have poor contacting with the solids, while the gas in the interstices has good contacting. This is less of a problem if the bubbles can be made smaller. Smaller particles result in smaller bubbles but also result in more entrainment and poor cyclone efficiency. However, a small amount of smaller particles (fines) added to the coarser particles produces bubbles nearly as small as if all the particles were small. The amount of fines required to achieve this effect is that needed to coat all the large particles with fines [80].

Schatz, Owen, and Snyder [229], in riser pilot-plant studies of dilute phase fluid transport, measured catalyst concentrations and

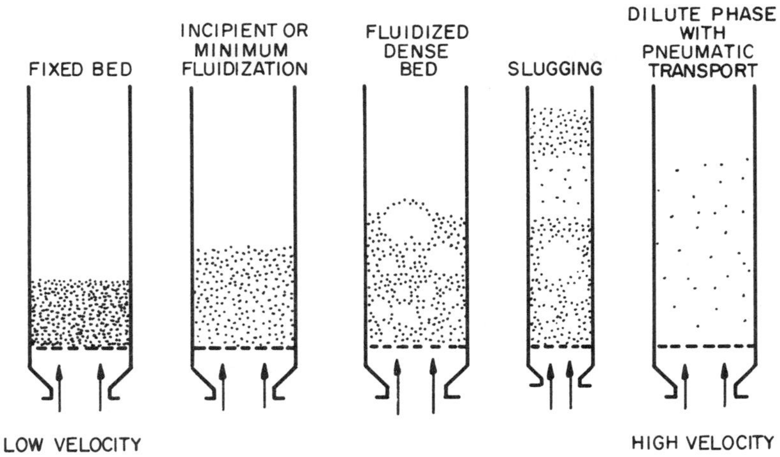

FIG. 18. The various states of an aerated bed. The gas velocity is increasing from left to right. If the vessel diameter is large enough, the slugging regime may not occur and increasing gas velocity will result in a smooth transition from dense bed to dilute phase. After Kunii and Levenspiel [228].

deviations from ideal plugflow for gas and catalyst. Saxton and Worley [63] conducted fundamental studies on the nature of fluidization in risers; they observed that local densities greater than 60 lb/ft^3 occurred near the riser walls, and that densities less than 3 lb/ft^3 occurred near the riser center line, indicating local velocities as much as twice the superficial vapor velocities of 35 to 75 ft/sec. General information on fluidization and pneumatic transport of solids can be found in Refs. 10, 80, 82, 84, 228, and 230.

2. Properties of Equilibrium FCC Catalysts

Table 8 gives a representative comparison of the properties of fresh and equilibrium zeolite and amorphous FCC catalysts as reported by Baker and Blazek [124]. The 0.20 to 0.25% carbon on the equilibrium catalysts is the residual level remaining after regeneration. Note the relatively low sodium levels and the sharply higher levels of iron, and the appearance of contaminant metals such as nickel and vanadium on the equilibrium catalysts. Bass [231] reports that clay-based catalysts surprisingly often have titanium

TABLE 8

Comparison of Properties of Fresh and Equilibrium FCC Catalysts[a]

	Amorphous high alumina		Zeolite XZ-2	
	Fresh	Equilibrium	Fresh	Equilibrium
Chemical analysis (dry basis), wt%				
Alumina[b]	28	21.5	31	25.4
Carbon	0	0.20	0	0.25
Sulfate	0.7	0.1	0.3	0.1
Sodium	0.03	0.02	0.04	0.04
Iron (ppm)	300	3900	700	3700
Metals (ppm of V + Ni + Cu)	0	162	0	259
Physical properties				
Surface area, m^2/g	415	140	335	97
Pore volume, cm^3/g	0.88	0.43	0.60	0.45
Average bulk density, g/cm^3	0.39	0.70	0.52	0.68
Particle size, micromesh sieves, wt%				
0 to 20 μm	2	0	2	0
0 to 40 μm	17	8	19	6
0 to 80 μm	68	69	75	75
Average particle size, μm	65	63	62	62
Catalytic activity				
Microactivity, vol% conversion	60[c]	59[c]	85	73
Carbon factor	1.0	1.1	0.6	0.6
Hydrogen factor	1.0	1.1	0.2	0.7

[a]Data from Ref. 124 for same commercial unit.

[b]Equilibrium values are lower due to presence of some residual low alumina of long age.

[c]After 24 hr/1050° F/60 psig, 100% steam.

inclusions in local concentrations far greater than levels indicated by bulk chemical analysis. Typical rare-earth element distributions and levels were discussed earlier. With the use of a scanning electron microscope/energy dispersive x-ray analyzer, Bass has also been able to distinguish the presence of plate-like kaolin clay, granular gel, and crystalline zeolite in the interior of lightly ground cracking catalyst powder [231].

Surface areas are sharply lower (by a factor of ~3 to 3.5) in equilibrium vs fresh catalyst; likewise, there is a drop in pore volume and an increase in bulk density. Letzsch [227] suggests that, in some cases, the initial loss in surface area with catalyst use may in large part be due to the collapse of small matrix pores and some zeolite regions which were not properly stabilized (i.e., had high residual sodium or inadequate rare-earth exchange levels).

For both amorphous and zeolite catalysts, the 0 to 20 μm fraction disappears and the 0 to 40 μm fraction is greatly lowered in the equilibrium catalysts compared to the fresh. This reflects the observation [62] that the finest part of the fresh catalyst added to a fluid unit is often lost on the first one or two cycles of operation, before it has much chance to do the work of cracking. Efficient regenerator cyclones will minimize fines loss and makeup required for catalyst activity maintenance. Beyond the immediate loss of fresh catalyst fines, most catalyst losses result from attrition to particles below the size range retained by the cyclones [62]; attritional losses are aggravated by excessive steam or air velocities. However, as indicated earlier, a distribution or gradation of particle sizes is required to maintain proper fluidization dynamics and cyclone efficiency [62]. Edgar [232] has demonstrated (from lab simulations and scanning electron-micrograph studies) that fresh FCC catalysts are not subject to thermal shock and fracture from rapid vaporization of volatiles when added to hot (1200° F+) regenerators.

Conversions are higher, and carbon producing (CPF) and hydrogen factors are lower for the zeolite, as expected. CPF is a useful guide to the degree of contamination of a catalyst, and is the ratio of coke yield for a given catalyst to that for a standard clean silica-alumina catalyst at the same conversion level [28]. The hydrogen factor—defined as the ratio of moles of H_2/moles of C_1+ moles of C_2's—increases with metals level.

Other data summaries on properties of fluid zeolitic catalysts have been reported [106, 107], as well as inspections for TCC bead catalysts [101, 105, 128, 131], silica-magnesia [128, 155, 174], etc.

3. Laboratory Evaluation of Cracking Catalysts

The activity of a catalyst can be defined as its capability of converting a chosen feedstock to specified product slate at a defined set of operating conditions [141]. Conversion-selectivity data acquired in laboratory pilot units at various temperatures and cracking severities (catalyst/oil ratios, residence times) are important bases in establishing the design of commercial units [12].

Although highly desirable, it is frequently impractical for a refiner to evaluate all new cracking catalysts after equilibration in a commercial FCC unit; it is thus necessary to resort to artificial laboratory-scale hydrothermal pretreatments with conditions of time, temperature, and steam partial pressure so chosen that deactivation to performance levels representative of commercial units is obtained [28, 106, 227, 233]. The deactivation is generally associated with losses in surface area, pore volume, and crystallinity [106, 227, 233]. While the primary measure of performance evaluation is the activity/selectivity pattern, as close a matching as possible of physical properties, including pore size distribution, is desirable [227]. Of course, any "masking effects" due to metals (or the coke derived therefrom) must be taken into consideration [28]. It should also be recalled that in commercial equilibrium catalysts there is a distribution of deactivations (due to the distribution in age) [227], while with lab-steamed compositions the samples are uniformly deactivated. Letzsch et al. [227] comment that for XZ-25 commercial cracking catalyst, treatment at 1520° F for 12 hr with 1 atm of 20 wt% steam (fixed fluidized bed) gave reliable deactivations, but that since catalyst compositions vary widely, no single set of steaming conditions will necessarily be optimum for all. Thus conditions encompassing the ranges of 980 to 1225° F and 5 to 120 psig of 100% steam have also been reported [233].

It is worthwhile to touch briefly on the variety of cracking test reactors that have been reported, with the understanding that such a broad and highly complex subject as laboratory reactor design constitutes a subject for review in its own right. The α-test, developed by Weisz and associates [101, 213, 234], involves cracking of n-hexane at 538° C and 9 sec catalyst residence time in a fixed-bed continuous-flow microreactor. This procedure avoids the potential testing variabilities encountered with complex gas oil feeds. Other microreactor systems have been discussed by Hall and MacIver [235], Chen and Lucki [236], and Nace [237].

Among fixed-bed cracking tests, the Davison Microactivity Test [106, 233] is particularly well known; after a specified thermal or hydrothermal pretreat of pelleted catalyst, reaction of 32.4 °API West Texas Devonian Gas Oil is effected at 900° F, ~ 5.8 catalyst/oil ratio with variation of space velocity via feed rate. The "D and L" [238], Cat A [151], Cat C [17, 18], and other [64, 239] fixed-bed tests have also been employed. A differential reactor utilizing a bed of fluid catalyst particles freed of fines has been developed [45].

Fixed fluidized beds have been employed extensively [107, 109, 110, 125, 127, 181, 240], including the Fluid Activity Index (FAI) test [114] for catalyst activity. Diverse circulating fluid bed systems [28, 129] and riser-type pilot plants [47, 99, 224, 241] have been reported. Detailed, more general discussions of laboratory reactor design have been published by Carberry and Butt [242], Doraiswamy and Tajbl [243], and Weekman [244].

Finally, in comparing catalytic cracking data from different literature sources, either lab scale or commercial, caution should be exercised to make sure that (1) catalyst residual carbon and metals levels and age are equivalent; (2) gasoline yields are based on the same 90% or endpoint (or that one is not comparing a yield based on an endpoint with one based on a 90% point, or one based on an ASTM distillation with one based on a True Boiling Point distillation; (3) process variables are really comparable (i.e., similar oil and inert diluent partial pressure, catalyst/oil residence times, riser vs backmixed dense bed, slip factors, feed preheat temperature, stripping, unit geometric configuration, etc.); (4) proper adjustments are made for the effect of temperature on reaction rates, particularly when comparing isothermal vs nonisothermal reactors, and for the heat-balance constraint for adiabatic (i.e., commercial) operation [12]; and (5) feedstocks are comparable, including the presence or absence of gasoline boiling range material.

VIII. DYNAMICS OF CATALYST, FEEDSTOCK, AND HARDWARE INTERACTIONS

A. Operating Variables vs the Constraints of the Operating "Window"

The objective of a modern short contact time FCC is frequently to maximize yields of high octane gasoline [74]. It does so in a highly flexible manner, operating within a framework of balances—heat, pressure, and chemical [141]. Its operations are complex (few variables can be changed independently) but operation is very stable because a change in one processing variable always results in an equilibrating response in others [141]. The unit thus shows a remarkable ability to accommodate operational variations. The FCC appears to be able to "thermally stabilize" itself automatically [12], a change in the coke-burning rate, for instance, through a series of changes in operating variables, will result in a compensating change in the coke-forming rate [245].

Although it may not be obvious to the casual observer, a commercial fluid catalytic cracker must function within fairly stringent boundaries. It is subject to enforced constraints from diverse sources, and its seemingly almost homeostatic normal interactions ("physiology") are occasionally perturbed by upsets ("pathology"). In addition to the restraints of feedstock variation, product specification, hardware/engineering, and catalyst represented in Fig. 2, unit performance is greatly affected by thermodynamics (from the aspects of heat balance and potentials in product selectivity), environmental factors, and, inexorably, by economics. Figure 19 represents some of these constraints—the "window" within which the chain of operations must occur. In this section, we will examine the nature of some of these complex interactions.

B. Effect of Operating Variables on Performance

1. Heat Balance as an Overall Frame of Reference

The adiabatic FCC heat balance derives from the concept of macroscopic energy balance of classical chemical engineering [245, 246]; i.e.,

$$\left\{\begin{array}{l}\text{Accumulation of}\\ \text{energy within}\\ \text{the system}\end{array}\right\} = \left\{\begin{array}{l}\text{transfer of energy}\\ \text{into system through}\\ \text{system boundary}\end{array}\right\} - \left\{\begin{array}{l}\text{transfer of energy out}\\ \text{of system through sys-}\\ \text{tem boundary}\end{array}\right\}$$

$$+ \left\{\begin{array}{l}\text{energy generation}\\ \text{within system}\end{array}\right\} - \left\{\begin{array}{l}\text{energy consumption}\\ \text{within system}\end{array}\right\}$$

To make a total energy balance, the first law of thermodynamics (conservation of energy) requires the last two terms to be zero, and for all steady-state operations there is (by definition) no accumulation. The equation may then be rewritten as

$$\left\{\begin{array}{l}\text{Transfer of total energy}\\ \text{into system through sys-}\\ \text{tem boundaries}\end{array}\right\} - \left\{\begin{array}{l}\text{transfer of total energy}\\ \text{out of system through}\\ \text{system boundaries}\end{array}\right\} = 0$$

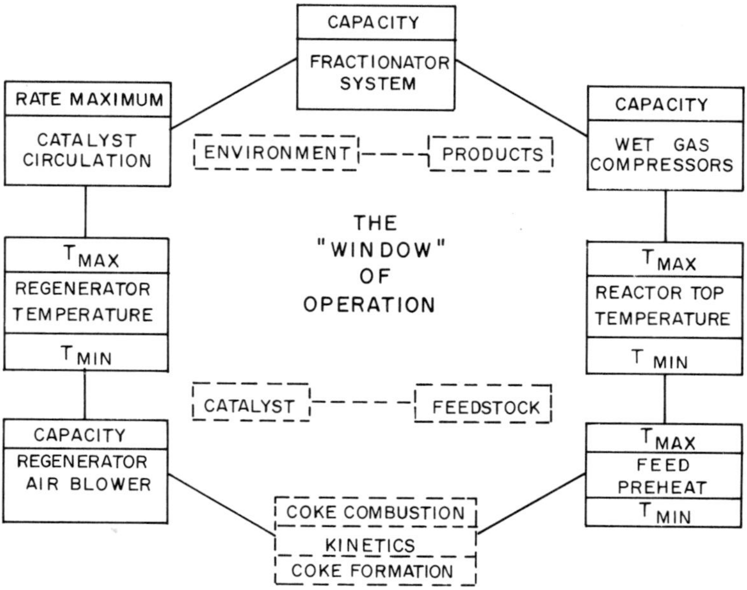

FIG. 19. Some of the major operational and engineering restraints imposed on a fluid catalytic cracker. Rates of coke formation and combustion of coke on catalyst are not classified as operating variables, but are shown because of their critical coupling role in the reactor-regenerator dynamics.

Neglecting the very small kinetic energy terms, this can be simply stated as the sum of the enthalpies of the reactants, minus the sum of the enthalpies of the products, minus the heat losses (by radiation and conduction) equals zero. Pohlenz [12] takes as a frame of reference the reactor outlet and the flue gas temperatures for the reactor and regenerator energy balances, respectively. This simplifies the calculation and directly produces equations where the heat of reaction terms are calculated at those respective reference temperatures [12].

Simplified expressions for the heat balance can, accordingly, be written [12, 69, 245]:

<u>Reactor</u>

$$-(CCR)(C_{Pc})(\Delta T) + (FF)(CFR)(\Delta H_{CF}) + (SR)(\Delta H_S) + (FF)(\Delta H_{RX})$$
$$+ (CKR)(\Delta H_{CS}) + L_{RX} = 0$$

Regenerator

$$(CCR)(C_{Pc})(\Delta T) + (CKR)(\Delta H_{CB}) + (AR)(C_{Pa})(T_{FG} - T_{AI})$$

$$- CKR(\Delta H_{CS}) + L_{RG} + CKR(\Delta H_c) - COR(\Delta H_{CO}) = 0$$

where
CCR = catalyst circulation rate, lb/hr
C_{Pc} = catalyst heat capacity, Btu/lb/°F
ΔT = temperature difference between regenerator dense bed and reactor dense bed/riser top, °F
CFR = combined feed ratio
ΔH_{CF} = enthalpy of combined feed at the reactor outlet temperature relative to the feed temperature, Btu/lb
SR = stripping steam rate, lb/hr
ΔH_S = enthalpy of stripping steam at reactor outlet temperature relative to steam inlet temperature, Btu/lb
FF = fresh feed rate, lb/hr
ΔH_{RX} = heat of reaction at reactor outlet temperature, Btu/lb
CKR = coke formation rate, lb/hr
ΔH_{CS} = heat of adsorption of coke at reactor outlet temperature, Btu/lb (exothermic, negative)
L_{RX} = heat losses from reactor by radiation and conduction, Btu/hr
ΔH_{CB} = coke heat of combustion at flue gas temperature, Btu/lb (exothermic, negative)
AR = combustion air rate, lb/hr
C_{Pa} = air mean heat capacity, Btu/lb/°F
T_{FG} = regeneration flue gas temperature, °F
T_{AI} = air inlet temperature, °F
L_{RG} = heat losses from regenerator by radiation and conduction, Btu/hr
ΔH_c = specific enthalpy of coke at flue gas temperature relative to reactor outlet temperature, Btu/lb
COR = CO rate in flue gas, lb/hr
ΔH_{CO} = heat of combustion of CO at the gas temperature, Btu/lb (exothermic, negative)

If these equations are solved for CCR and equated, a relationship expressing coke rate as a function of the independent operating variables may be derived [12].

As schematically depicted in Fig. 20, the reactor zone is physically and operationally coupled to the regenerator [245]. This can

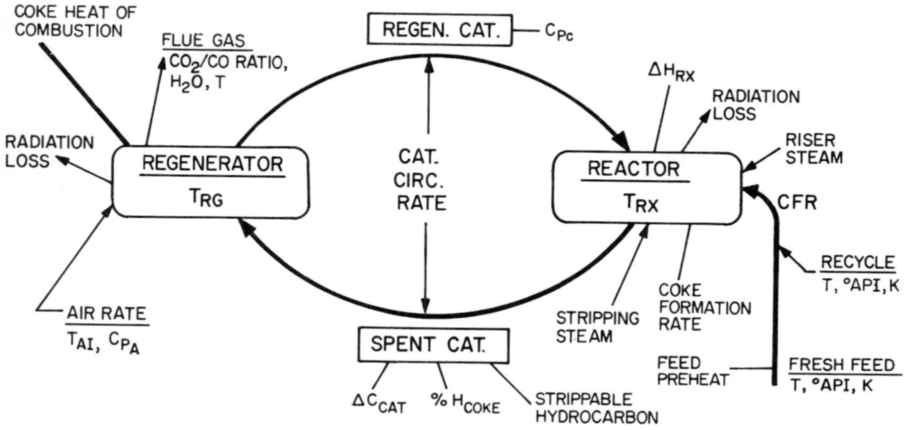

FIG. 20. Visualization of major elements impacting on heat balance in fluid catalytic cracking. Note the central role of the catalyst circulation rate in communicating between reactor and regenerator. Interactions assembled from Refs. 12, 67, 69, 74, and 245.

be placed in the perspective of the hardware configuration by reference to Fig. 8. The unit is "heat-balanced" in the sense that the heat for vaporizing and reacting the feed is derived by combustion of the produced coke in the regenerator, and the released heat is transferred by the catalyst circulation to the reactor [69, 141, 245]. An overall energy balance shows that the energy released by the combustion of coke becomes the sensible and latent heat of the reactor effluent and flue gas [12]. Expressed another way, the coke yield in an adiabatic FCC is essentially that required to satisfy the heat load [67, 74]; above and beyond feed preheat, commercial cracking units (excluding the distillation column) using conventional gas oil feeds generally don't require heat exchange with external heat sources or sinks [12]. Heats of reaction are very small [12], and a calculation method (based on the difference in heats of formation of feedstock and reaction products from their elements) has been reported [247].

The amount of coke burned may be calculated from the air rate and flue gas analysis, with the coke hydrogen content determined by oxygen disappearance [12]. If one considers thermochemical data, it is evident that the heat released from oxidation of CO to CO_2 is about 2.3 to 2.6 times greater/lb of carbon than that from combustion of carbon to CO:

Heat Released, Btu/lb Carbon [240]	
$C + \frac{1}{2}O_2 \rightarrow CO$	4,440
$CO + \frac{1}{2}O_2 \rightarrow CO_2$	10,160
$C + O_2 \rightarrow CO_2$	14,600

It also follows that not only is the CO_2/CO ratio an important variable in calculating the quantity of heat release from coke combustion [81, 240, 245], but also the coke hydrogen content [12, 245]. Himmelblau [246] cites a value of -57.8 kcal/g-mole for ΔH formation of water, i.e., $H_2(g) + \frac{1}{2}O_2(g) \rightarrow H_2O(g)$. Although the hydrogen in coke is chemically combined with carbon (Pohlenz [12] cites a range of 7.6 to 11.0 wt% H), reasonable thermochemical calculations may be made assuming that coke is a physical mixture of hydrogen and carbon [12]. The amount of (combined) hydrogen burned in the regenerator is directly related to efficiency of stripping [245], where entrained hydrocarbon vapors [69] must be removed as completely as possible [65].

A recycle stream can, in principle, play an important role in heat removal from the system [12, 69, 141], since it enters as a liquid, absorbs latent and sensible heat, and leaves as a vapor. However, with the advent of short contact time riser cracking and high activity zeolite catalysts, there is generally little recycle (i.e., 5% or less) in modern FCC units [67, 74, 94]. (See Section VI-B-2).

Use of feed preheat—addition of thermal energy to the front end of the process—can be used to supply a portion of the heat needed by the reactor system [12, 39, 69, 73, 74, 141, 240], and is particularly useful in utilizing the low coke-forming potentialities of zeolite catalysts [65]. This is commonly effected via stepwise heat-exchange with hot product streams and/or a fired preheater [65, 141]. This represents a tradeoff where there is reduction of heat supplied by hot catalyst circulation (CCR)—from combustion of product coke in the regenerator—at the expense of fuel burned in the preheater [12, 69, 141]. It also follows that, for a given reactor outlet temperature, every unit will have an economical maximum feed preheat temperature, above which the catalyst circulation (and therefore the conversion) will become too low, requiring use of a heat removal system for the regenerator [69]. (Note that in the direct cracking of atmospheric residua (reduced crude), the quantity of heat evolved from burning the large amount of coke formed is greater than the heat removal capacity by flue gas and CCR, and regenerator design includes steam-generating coils around the vessel circumference to remove the excess [72, 121].

The chemical energy of the flue gas in the regenerator effluent has often been recovered at the back end via an external CO-boiler/power recovery system (Section V-B) [39, 78, 88]. However, the CO unoxidized in the regenerator also represents a significant loss of energy from the FCC process [240]. Advances have been made in achieving complete combustion of CO by modifying mechanical design and operating philosophy in the regenerator [39, 78, 79, 249], but CO-conversion has been more successful using controlled combustion catalysts [20, 81, 240] which allow complete CO oxidation during regeneration and have potentially increased the total heat release from coke combustion by over 50% [240]. Thus coke requirements for a given set of conditions may be reduced [39]. (With conventional FCC catalysts, about half of the carbon burned in the regenerator appears as flue gas CO, the CO_2/CO ratios typically being about one [240]). With hydrotreated feeds (see Section VI-B-2), the constraints of heat balance may occasionally influence the degree to which coke may be lowered (and hence the severity of hydrotreating) before requiring offsetting heat inputs by feed preheat, etc. [47, 59].

2. The Key Reaction Variables

The most comprehensive discussions of reaction variables can be found in the classic paper of Pohlenz [12] and in a recent overview by Montgomery [141]. The major independent reaction variables [12, 141] in a commercial FCC operating in the maximum gasoline mode are:

a. Reactor temperature
b. Combined feed rate (the sum of fresh feed and recycle rates) which controls the catalyst/oil residence time
c. Combined feed (feed preheat) temperature
d. Reactor pressure
e. Catalyst activity

These are directly regulated, usually with control devices [12, 141] (see Sections V-A and V-B). Dependent variables, on the other hand, are responsive to changes in the independent variables [12, 141]. These include:

a. The catalyst circulation rate, CCR (and hence the catalyst/oil [C/O] ratio)
b. The regenerator temperature
c. Conversion

The latter is subject to the influence by many independent variables. The air rate required to support combustion of coke deposits may also be classified as a dependent variable [141]. Of course, feedstock variations will profoundly affect all variables.

In contrast, with most pilot plants, variables can be changed independently of heat balance [12, 138, 141]. Here reactor and regenerator temperature may both be independently controlled, and greater flexibility for experimental work is allowed, particularly in that the C/O ratio (CCR) can be used as an independent operating variable to control reaction severity [12, 138].

Independent Variables

a. Reactor Temperature. Catalytic cracking at higher temperatures—and short contact times—is generally associated with higher gasoline octane numbers [66, 67, 74, 77, 94, 99, 141, 250, 251] and greater production of light olefins, especially C_3 and C_4 olefins [65, 66, 94, 99, 141, 226, 250, 251]—with their potential for alkylation feed [97, 252]. Cracked gasolines from high-temperature operations are particularly useful blending components for lead-free gasoline [94, 226]. With short contact time riser cracking, the improvement in octane can be achieved without appreciable overcracking of gasoline [19, 65] and levels of 92 to 97 R + O have been reported [67, 74] for riser outlet temperatures of 980 to 1020° F in commercial units. There appears to be less effect on motor octane, M + O [226]. Peak gasoline yields (i.e., maxima in the plots of gasoline yield vs conversion) are sometimes observed at lower conversions as the temperature is raised [99, 250, 253]. The incremental octane improvement tends to level out as temperature is increased [74], however, and an optimum exists for every operation and feedstock.

The characteristic shifts in product distribution with increasing temperature (at constant conversion) are shown for the riser FCC pilot plant cracking of 28.4 °API gas oil [99] in Fig. 21 (see also Table 3). Note that as reactor temperature increases, dry gas (C_3^-) and total C_4's increase, but gasoline and coke decrease. Figures

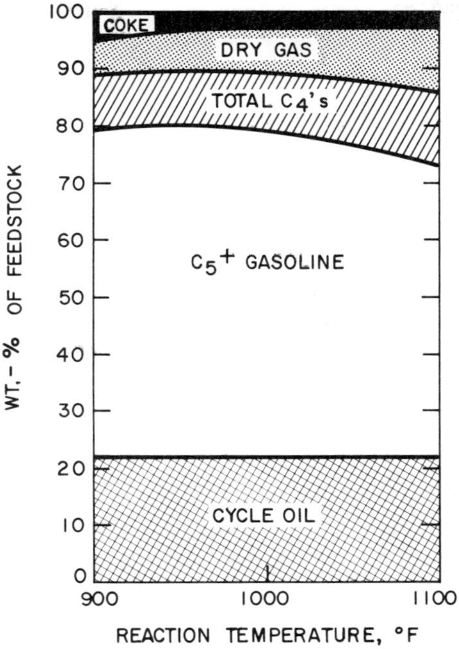

FIG. 21. Effect of reaction temperature on product distribution at 78 wt% conversion in short contact time riser FCC pilot-plant cracking of hydrotreated heavy vacuum gas oil over commercial zeolite catalyst. Adapted from Owen, Snyder, and Venuto [99].

22(A), 22(B), and 23 show trends in composition of dry gas, total C_4's, and C_5^+ gasoline with increasing temperature. There are increases in olefin yields of every fraction; these are particularly sharp in the C_3 to C_5 range. This reflects, in part, the influence of lower hydrogen transfer rates relative to cracking rates at the higher temperatures [99, 254]. As is evident in Fig. 22(B), the ratio of iso-C H / n-C_4H_{10} also decreases with increasing temperature, reflecting equilibrium factors [11].

Strikingly similar product shifts with increasing temperature were observed by Olsen and Sterba [250] in an earlier pilot-plant FCC study of the cracking of 31.5 °API Mid-Continent gas oil over low-alumina synthetic silica-alumina catalyst at 800 to 950° F. Olsen and Sterba [250] also noticed an increase in both motor and research clear octane number and in octane sensitivity. There was

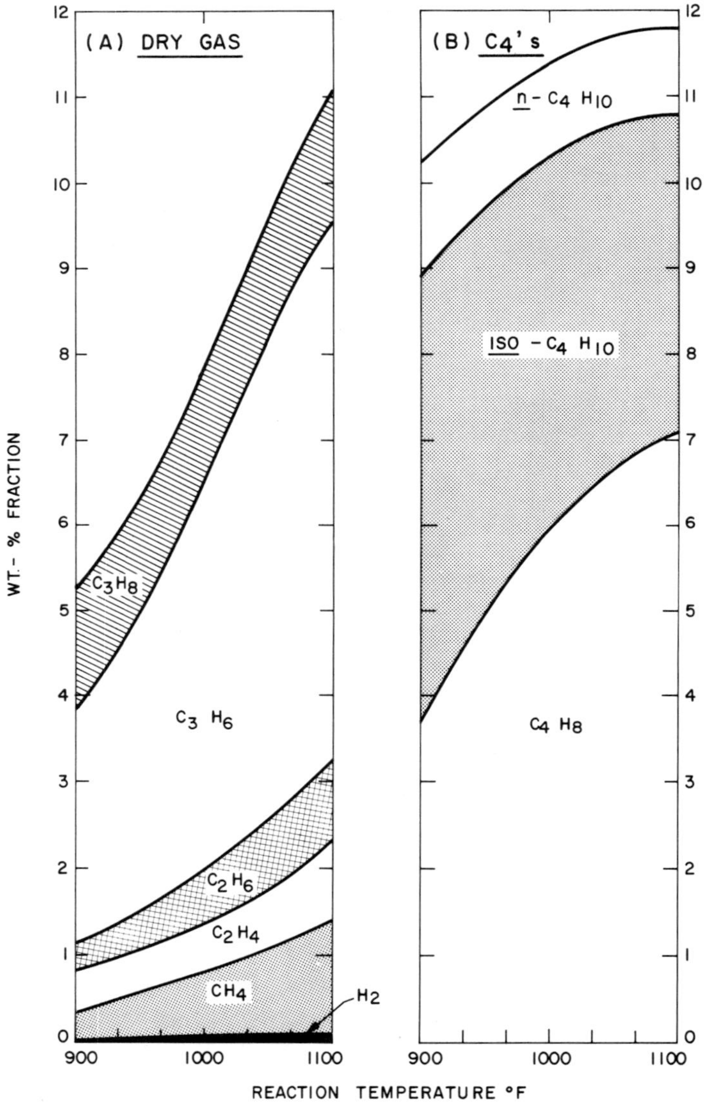

FIG. 22. Effect of reaction temperature on proportions of components in (A) dry gas and (B) total C_4-fraction at constant conversion (78 wt%) in short contact time riser FCC pilot-plant cracking of heavy vacuum gas oil over commercial zeolite catalyst. Adapted from Owen, Snyder, and Venuto [99].

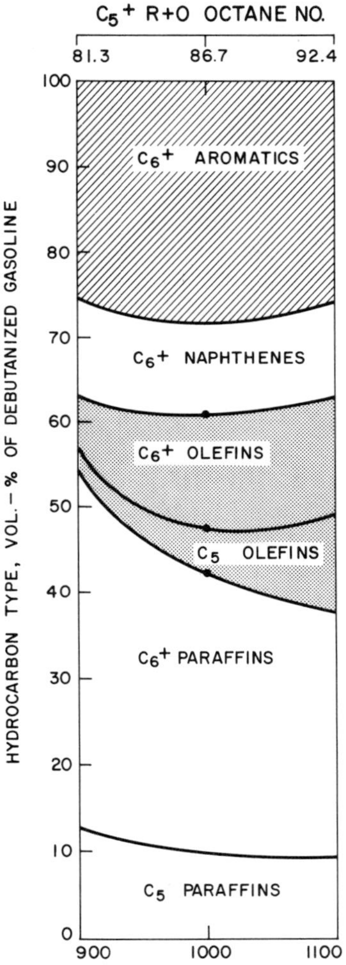

FIG. 23. Trends in octane number and yields of major hydrocarbon types in C_5-356° F at 90% A. S. T. M. gasoline with increasing reaction temperature (80 vol% conversion). Estimated from data of Owen, Snyder, and Venuto [99].

also a decrease in lead susceptibility and an increase in gasoline sulfur content in the gasolines obtained at higher temperatures. As discussed by Montgomery [141], however, the changes observed in a commercial FCC with increased reactor temperature may differ greatly from those for a carefully segregated temperature variable study in a pilot plant, due to the complex coupling of the operating variables in the former.

b. Combined Feed Rate. The central importance of combined feed rate—the sum of fresh feed and recycle rates—to heat balance [12] was emphasized earlier. Before the advent of zeolites, high recycle ratios compensated for the poor selectivity. First pass conversions were low, with recycle essentially allowing longer catalyst/oil contact times [141] for the unconverted material. (See Section VI-B-2.) This of course had the disadvantage of lowering fresh feed capacity [12, 141]. Montgomery [141] has discussed the effect of increasing combined feed ratio (at constant conversion) upon yield structure in pilot-plant cracking over amorphous catalyst.

c. Combined Feed (Feed Preheat) Temperature. The heat balance relationships emphasize the importance of a properly adjusted fresh feed temperature (Fig. 20) and combined feed ratio [12]. Use of feed preheat (FPH) allows a reduction of C/O ratio [12, 69, 73, 240], which ultimately shows up as decreased coke production [12, 69, 73, 240], better selectivity to liquid products [12, 73], and increased unit capacity [69]. Combined feed temperatures may range up to 700 to 750° F [73], with feed vaporization up to 75% or higher occurring in many fluid units [69] (depending on feed boiling range); of course thermal cracking—which may lead to furnace coking and fouling—must be avoided [65, 69]. Again, the heat balance (and perhaps also viscosity/pour point considerations) dictates the minimum feed preheat temperature.

d. Catalyst/Oil Residence Time. Experience has shown that the most efficient way to use high activity zeolite catalysts is with short catalyst/oil contact times [65, 67, 73, 74, 94, 141]. Exposure of a few (1 to 4) seconds in riser cracking is commonly all that is required to effect nearly complete conversion of the nonaromatic part of a feed [65], although this may vary with feedstock characteristics and yield objectives [74]. At short contact times, selectivity for gasoline yield is maximized since less opportunity for secondary cracking of gasoline (over-cracking) is permitted [67, 73, 74, 94, 141]. There is also less coke formation [73, 141] and preservation of C_3 and C_4 olefins (for alkylation) that might otherwise be drained off via hydrogen-transfer reactions [74].

e. Pressure. Although it is, in fact, an independent variable, reactor pressure is often set by unit capacity and main column or gas compressor limitations [12]. Pohlenz [12] also notes that pressure has little effect on conversion or selectivity over the ranges commonly employed.

f. Catalyst Activity. The type and activity of FCC catalyst is of great importance (see Section VII-D); quality and activity must be maintained by an appropriate rate of addition of fresh catalyst make-up [12, 141], and are a direct reflection of the fraction of total inventory replaced per day [66].

Dependent Variables

In moving to a discussion of dependent operating variables, it is worth emphasizing that in units functioning under controls such as those described in Sections V-A and V-B, catalyst circulation rate (CCR) and C/O ratio are not subject to direct control, and may be determined from the heat balance over either the reactor or the regenerator section [12]. Of central importance are the mutually-dependent interactions between feed preheat (FPH), CCR, and regenerator temperature (T_{RG}) [12, 141, 240]. We will illustrate these principles for several possible situations and from several viewpoints.

Consider a situation where CCR is regulated by temperature control in the reactor and all other independent variables are kept constant [12, 141]. Let us increase the level of FPH ("front end"). At constant reactor temperature, the controller "calls for" less hot, regenerated catalyst, and CCR (and C/O ratio) will decrease. The drop in C/O severity will result in lower conversion [75]—and also coke-make [12]—and increased yield of liquid products usually results [67, 74, 141]. Although the lower coke formation rate might be expected to decrease the temperature of the regenerator (T_{RG}) (because of less combustion required), in most cases, T_{RG} will rise [12, 141, 240], reflecting the dominance of the lowered rate of heat removal from regenerator to reactor because of the slower CCR [240]. This is often expressed as an increase in the ΔT (regenerator - reactor) [12].

Sometimes T_{RG} limitations (metallurgy), particularly in older units, impose an upper limit on FPH temperature [141, 240]. Note also that a sudden loss of recycle, i.e., a pump failure which prevents influx of a cooling stream, would simulate directionally an increase of FPH [12]. Of course, the CO_2/CO ratio in flue gas will also affect CCR, and use of controlled combustion catalysts may reduce or eliminate thermal requirements for FPH [39, 69].

The converse of the above sequence of interactions also occurs, i.e., as FPH is decreased, ΔT (regenerator - reactor) decreases and coke-make increases [12]; a similar pattern occurs if catalyst cooling or spray water addition to regenerator is practiced rather than lowering FPH [12]. As phrased by Pohlenz, the system acts as a "flywheel" in which "coke rate opposes the direction in which the regenerator temperature moves" [12]. The T_{RG} is thus stabilized to large fluctuations in feedstocks and operating conditions. In another sense the objective is to balance FPH with the coke-forming tendencies of the feedstock to achieve a high T_{RG}. Generally, higher FPH is required with better quality feedstocks [67, 74].

The lower coke rate obtained with higher FPH is also reinforced by an increase in efficiency of stripping [12, 141]: with the decline in CCR, not only is there less entrained hydrocarbon vapor entering the stripper, but also catalyst residence time in the stripping zone is prolonged. Thus more of the hydrocarbon vapor entrained in the interstices and void volumes of the catalyst particles is swept upward to join the reactor effluent and less hydrogen-rich material is carried from the reactor for combustion with (nonvolatile) coke-on-catalyst [75]. The lower CCR may also be associated with less attrition loss and lower quantities of CO_2 (entrained with the regenerated catalyst) that could ultimately cause scaling problems (as carbonate) in downstream gas plants.

High regenerator temperatures may, in fact, be desirable to obtain low levels of carbon on regenerated catalyst [67, 74, 78]. Note that use of controlled combustion catalysts to promote CO-burning may also increase T_{Reg} (dense bed) typically by about 100° F [240], which increases the coke-burning rate severalfold. Air rate is, of course, dependent upon the amount of coke produced, and hence severity of operation [141]. In conventional regeneration, no more air than that required to produce a detectable oxygen breakthrough is allowed into the regenerator bed; insufficient air leads to buildup of carbon on regenerated catalyst, while an excess of air leads to uncontrolled CO combustion—afterburning [141]. The ΔT between regenerator dense and dilute phase or dense phase and flue gas is a sensitive indicator of oxygen level at the boundary of the dense phase [12]. This ΔT is sometimes automatically controlled by venting ("snorting") an appropriate fraction of the blower air [12, 141].

3. Optimization for Light Fuel Oil

In North America, demand for middle distillate (diesel, home heating oil, turbine fuels, etc.) is mainly satisfied by using straight run materials, with some contribution from FCC light cycle oil (LCO) [138] (see Fig. 1); further, adjustment of gasoline/LCO cut point can generally meet the seasonal swings in demand for fuel oil. In other areas (including Europe), demand for heating oil exceeds the quantities naturally present in crude (Fig. 1), and FCC operation is geared, for the most part, toward LCO optimization [138]. Demand for light fuel oil, then, may be viewed as a product constraint on FCC operation, and operational response to this constraint again demonstrates the extreme flexibility of this molecular weight reducing refining tool [74].

Whittington, Murphy, and Lutz [74] and Murphy [65] emphasize the need for relatively mild, low-severity first-pass conditions to avoid recracking and to maximize middle distillate yields. This can be achieved by using a higher combined feed ratio (recycle of heavy cycle oil) [65, 74, 94, 138] and simultaneously lowering conversion [138]. The presence of heavy cycle oil reduces the partial pressure of the LCO, thereby lowering its recracking rate [74]. Segregated cracking of fresh feed and recycle has occasionally been practiced [74, 92] but utilization of a common riser is preferred for improved yields [74]. Sometimes a higher IBP feed is also used [64]. Reactor temperature may be decreased [94, 138], but the extent of lowering may be limited by an octane constraint or the fraction of higher boiling components in feed [138]. Thus conversion must generally be reduced by lowering the number of catalytically active sites in the reaction zone [138]. Pohlenz [138] notes that, if we assume a measured catalytic activity is proportional to the number of sites/unit mass, the number of active sites in a riser is related to the product of (activity) X (CCR). Thus decreased catalyst activity [74, 138] and increased combined feed temperature (FPH) [74, 138] (to lower CCR and C/O ratio and also provide heat for vaporization of recycle) may be required to help reduce first-pass severity. Earlier, with amorphous catalysts, operation for light fuel oil optimization was sometimes conducted at 55 to 60 vol% conversion [64, 94].

Lower first-pass severity also results in the incomplete cracking of heavier fractions [12, 65, 74], and these must be recycled [12, 65, 74, 94] to obtain high light fuel oil yields. Hence residence time must be controlled and balanced against quantity of recycle [138]. Proper FCC operation can thus generate significant quantities of middle distillates which, with moderate hydrotreating, can also meet specifications of Jet-A turbine fuel [138].

4. Aviation Gasoline

During World War II there was great demand for high octane aviation gasoline [251]. By both single-stage high-temperature operation (which simultaneously maximized both aromatic base aviation stocks and raw materials for alkylation) and two-stage (gasoline recycle/recracking) operation, FCC was adapted to meet another product constraint [251].

C. Dynamics of Conversion in Riser/Reactor Zone

1. Phenomena of Change during Riser Cracking

Several important fluid cracking parameters are plotted vs residence time in riser (or vertical distance up the transfer line) in Fig. 24. From the instant of contact between catalyst and oil vapors at the riser base, all subsequent interactions are complicated because conditions are constantly changing throughout the length [74]. Concomitant with conversion of gas oil are a significant molar expansion coupled with acceleration of both vapor and catalyst [10, 12, 74, 255], deposition of coke and impurities on the catalyst [62, 74], and a decay of catalyst activity and selectivity [75, 109, 110, 256].

The Mixing Zone

In order to fully utilize the intrinsic catalytic activity of zeolites, proper design to allow intimate mixing at the point of initial contact of catalyst and oil is critical [19, 61, 64, 76, 116]. Further, there should be even distribution of feed with minimum back-mixing (near-plug flow) during the cocurrent flow of catalyst and vapor through the vertical riser tube [61, 62, 64, 74, 116]. Injection of nearly vaporized feed (high feed preheat) [64, 76], rapid vaporization of feed in riser [64], increased atomization and dispersion of liquid feed [64], use of vertically oriented feed nozzles [74], multiple injection points or feed nozzles [63, 74, 118], and use of dispersion steam (especially with residuum or reduced crude where lowering of hydrocarbon partial pressure is desired [72, 74] have all been reported as aiding efficient catalyst/oil mixing—and improving conversion [64].

Acceleration, Slip, and Catalyst Distribution

With initial catalyst/oil contact, fluidization occurs [257], and the rapid molar expansion of cracking causes a sharp increase in vapor velocity, with acceleration of catalyst over a few seconds from relatively low initial velocity to one approaching that of vapor [12, 74, 229]. Although the gas velocity tends to drive individual catalyst particles upward, there are the opposing effects of gravity and inertia [10], with the result that the solid velocity is less than the gas velocity. This difference is known as the "slip velocity" [10]. For any section of the riser, the slip ratio S can be defined [74] as the ratio of catalyst residence time/vapor residence time, i.e.:

$$S = \frac{\tau_c}{\tau_v} = \frac{V_v}{V_c}$$

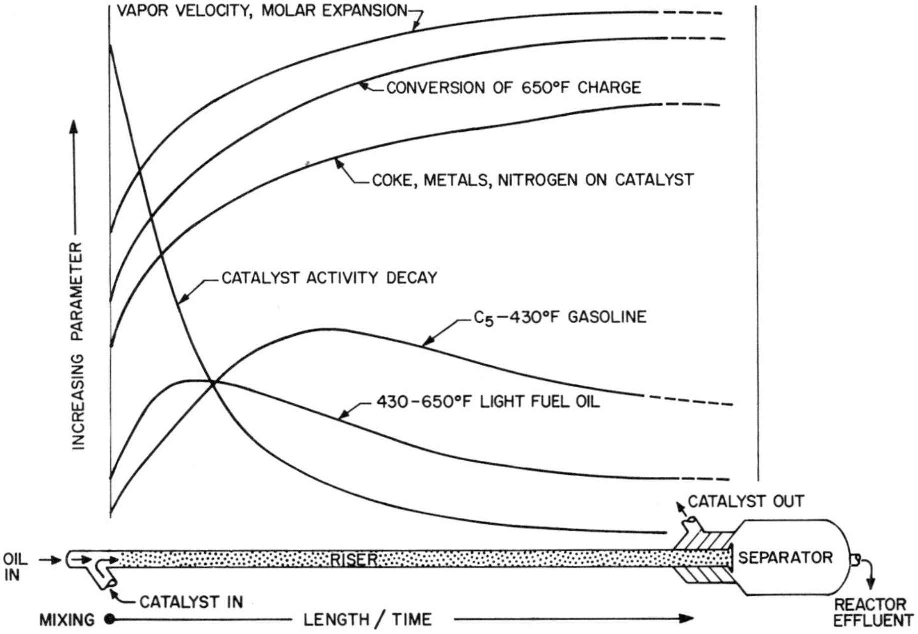

FIG. 24. Representative profiles and trends with riser length/catalyst oil contact time for some key parameters in fluid catalytic cracking of a 650° F+ gas oil. Curves are intended to show basic directional patterns only, and do not necessarily map kinetic trajectories. Schematic is adapted from data in Refs. 74, 138, and 255.

where τ_c = catalyst residence time
τ_v = vapor residence time
V_v = vapor velocity
V_c = catalyst velocity

Note that after the initial acceleration, catalyst velocity differs from the vapor velocity by an amount approaching the free fall velocity V_f [74], and the following equation holds:

$$S = \frac{V_v}{V_c} = \frac{V_v}{V_v - V_f}$$

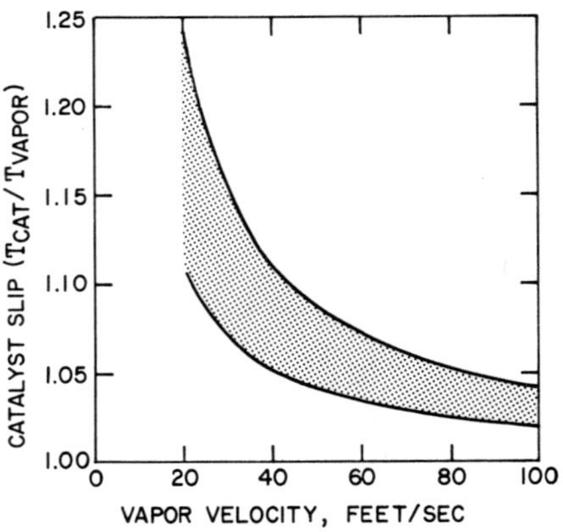

FIG. 25. Plot of catalyst slip (after initial acceleration zone) vs vapor velocity in commercial riser cracking. After Whittington, Murphy, and Lutz [74].

As shown in Fig. 25, slip can be significantly reduced by high (final) vapor velocities, say above 80 ft/sec [74, 255], and may approach about 1.1 near the top of the riser [255]. Slip over an entire transfer line, however, has been reported to average as high as 1.5 [255]. Catalyst holdup in any section of the riser may be expressed [255] as

$$dH = \frac{CCR \times S \times dL}{60 \times V}$$

where H = catalyst holdup, tons
CCR = catalyst circulation rate, tons/min
S = slip factor, vapor velocity/catalyst velocity
L = length, ft
V = vapor velocity, ft/sec

High vapor velocities not only reduce the catalyst holdup that occurs in the riser bottom, but also allow considerable improvement in catalyst distribution [62, 74, 229]. Radial maldistribution (i.e., high localized catalyst concentrations at the wall) and other aberrations that cause deviations from ideal plug flow are also minimized with high vapor velocities [63, 74, 229], with catalyst distribution tending to be more uniform with extent of vertical displacement up the riser [63] (see Section VII-E-1). Velocities up to 70 ft/sec have been used commercially with good results [74]. Gas residence time τ_V has been directly measured in a pilot riser using radioactive krypton tracer [129].

Mechanical and Circulation Effects

From a mechanical standpoint, a riser/reactor must be designed to cope with thermal expansion, pressure forces, erosion, corrosion, and heat flow limitations [63]. One of the negative aspects of very high vapor velocities V_V is erosion of equipment and attrition of catalyst [74] (see Section VII-E-2). By use of erosion-resistant materials of construction and proper flowing velocity range, the above problems can be minimized [62, 116] and long times (on stream) may be attained. When efficient multistage cyclones are employed in the reactor, the catalyst escaping the second-stage cyclones is very fine, and not worth returning via slurry [94].

When high sulfur feeds are used, however, metal loss in the area of the reactor grid holes may be quite severe from a combination of erosion and corrosion [63].

Catalyst circulation is maintained by the differences in density of the fluidized catalyst at various points in the flow circuit, with density, in turn, being determined largely by the velocity of the fluidizing gas [258]; typical densities are: reactor, regenerator beds, 25 to 35 lb/ft^3; catalyst standpipes, 35 to 45 lb/ft^3; riser and lift lines, 2 to 5 lb/ft^3 [258]. It is because the high densities in the beds and standpipes (compared to those in the riser length) essentially generate a hydrostatic head that circulation is possible (see Section VII-E-1). The pressure head (ΔP), which must be sufficient to overcome losses due to friction and the pressure drop across the various slide valves and grids, is the predominant factor in determining unit (riser) height, which is typically about 150 ft for staged units [258]. It should also be remembered that, for every unit, there is a physical limit to the catalyst circulation rate, i.e., a maximum in terms of tons/min.

Pressure and Temperature Effects

There are a number of benefits from higher (reactor/regenerator) pressures—improved regeneration kinetics (see Section VIII-F-2), reduced vapor load (velocities) in the fractionator, lower (wet gas) compressor horsepower requirements, improved efficiency of power recovery systems, etc. [74]. But there are also some counterbalancing adverse effects: an increase in carryover of inerts with the regenerated catalyst to the reactor—and ultimately to the gas recovery system—can occur [66, 74], as well as an increase in catalyst entrainment at the same velocity. Since the regenerator pressure is coupled to that of the reactor system [12], air blower horsepower requirements will increase and there will be an increased pressure drop over the regenerator flue gas orifice chamber [74]. As oil partial pressure increases, the rate of coke deposition on circulating catalyst may increase significantly [66, 74], although this may be partially countered through the use of diluents (such as inert gases and steam) [66, 72, 74]. Of course, the use of inert diluents will tend to nullify the advantages of reduced fractionator loading and reduced compressor horsepower requirements.

Maximum reactor temperature (Fig. 19) is set by materials of construction (metallurgy) or by possible increasing impact of undesirable side reactions. For very high reactor temperatures, the metallurgy constraint can in part be alleviated by using low alloy or internal insulation instead of hot wall carbon steel [74].

2. The Impact of Catalyst/Oil Contact Time on Selectivity: Riser vs Dense Bed Cracking

As depicted in Fig. 24, after the finite time of cocurrent vertical ascent through the riser line, it is necessary to effect disengagement of catalyst from the gaseous products [74]. Passing the riser effluent into a low-velocity (conical or wider diameter) vessel has commonly been used for an initial separation [74]. A few percent of incremental conversion may actually be gained by such a design [70, 74] or by employing a reactor catalyst level that can essentially be classified as a dense bed.

But these types of catalyst design result in contact of the hydrocarbon vapor with a high density of partially deactivated catalyst, allowing extremely high, localized catalyst/oil (C/O) ratios [70]. With the back-mixing that inevitably occurs when catalyst is held up in a dense bed environment, secondary or side reactions occur, generally with the formation of less valuable gas and coke at the expense of gasoline [19, 50, 61, 62, 65, 66, 73, 94, 143, 226]. Thus, in most modern units using high activity zeolite catalysts, not only is dense-phase contact of catalyst and oil minimized beyond the interval of cocurrent flow through the riser [19, 50, 61, 73, 74, 92, 101, 141, 226] (i.e., by some "quick quench" [256] such as passing vapors directly into the cyclones [74]), but contact time in the riser itself is kept short, i.e., less than 5 sec in some cases [66, 256] (see Section VIII-B-2). In fact, Vermilion [66], in a study where a series of samples was analyzed along the vertical path of a commercial riser, showed that in only 2 sec of C/O contact time, almost 60% conversion occurred with a zeolite catalyst, and that optimum conversion was obtained at 4 sec, with relatively little added in a subsequent dense bed stage.

This minimizes exposure of hydrocarbon to highly coked—and therefore less selective—catalyst [74]. This is particularly important at high temperatures [74] (see Section VIII-B-2). Note the maxima in gasoline selectivity (and light fuel oil) at relatively short contact times as shown schematically in Fig. 24.

In bed cracking with back-mixing, the coke content of the catalyst is approximately that of the spent catalyst [62]. In riser cracking, however, at any point the coke on catalyst is only that laid down by the cracking which has occurred up to that point [62]; i.e., each successive increment of cracking takes place on a catalyst surface that has the lowest possible carbon deposit and highest activity for a given C/O ratio [74]. Thus a riser system maintains the activity and selectivity of the catalyst during a greater portion of the cracking reaction [62]. The need for short contact time (i.e., riser) cracking is particularly critical for zeolite catalysts as contrasted to amorphous, since with the latter, coke does not build up so rapidly [94]; in fact, the low activities of the amorphous materials made it necessary to recycle [94], and also the selectivity of the amorphous-type catalysts is less affected by coke buildup [62, 77, 94, 240, 249].

3. Restraints Arising from Coke Deposits and Metals Poisoning

Effect of Residual Catalyst Carbon on Activity and Selectivity

One of the most important characteristics of zeolite-containing FCC catalysts is their susceptibility to loss of activity [75, 94, 129, 240, 249] and selectivity [20, 64, 75, 78, 94, 129, 150, 185, 226, 256] as coke on catalyst (C_{cat}) increases. In this regard, zeolites, in short contact time riser operation, appear to be more markedly affected by residual catalyst carbon level than conventional amorphous materials [62, 77, 94, 240, 249]. For this reason there has been a driving force for more efficient regeneration (vide infra) [77, 129, 256], with levels in the range of 0.1 to 0.25 wt% carbon on regenerated catalyst commonly being cited as desirable [64, 65, 101, 116, 143, 150, 256] and at levels as low as 0.05 wt% in several references [78, 249].

The selectivity advantages with low C_{cat} most commonly cited are higher gasoline yield [64, 78, 94, 129] and lower coke and gas yields [78, 94, 129, 150]. The magnitude of the effect of C_{cat} upon cracking patterns and yield distributions (probably with minimal distortion from metals effects) can be seen in some data of Wachtel et al. [129] in Fig. 26. As C_{cat} increases from 0.1 through 0.4 wt%, note the decline in conversion (with Δ conversion/ΔC_{cat} response of about 3.2 vol%/0.1% increase in C_{cat} over this interval), and the increase in coke, C_3 and lighter gas, and butanes at the expense of C_5+ gasoline. These changes in product distribution are in the direction of less selective cracking, i.e., that of silica-alumina (see Tables 6 and 7); they are perceptible with even small ΔC_{cat}, with the specific magnitudes of response dependent upon feedstock and catalyst [129].

Wachtel et al. [129] attribute the loss in gasoline selectivity with increasing C_{cat} to blockage of zeolite sites by carbon; others [226, 249] have suggested that, with low levels of C_{cat}, there is better accessibility and minimal coke-induced diffusional limitations within the catalyst. Ritter [150], based on pilot-plant studies, suggests that most of the coke is preferentially deposited in and around the active cracking sites of the zeolite particles; since most of the cracking is assumed to occur at the zeolite sites, one would then expect a change in selectivity (as well as activity). He also observes [150] that the coke on the zeolite should be the last removed during regeneration, and that essentially complete coke removal is desirable to maximize the potentials of the zeolite. A further consequence of this line of reasoning is that attempting to offset the negative effects of increased C_{cat} by increased process severity would be futile, since trends toward undesirable products would be accelerated [129, 150].

Effects of Metals Contaminants on Activity and Selectivity

For discussions on metals relating to origins, feedstocks, and catalyst, see Sections III-B-2, VI-A, and VII-D-2, respectively. Any metals poisons in the FCC feed, even very small concentrations, will deposit almost quantitatively on the cracking catalyst. These can potentially accumulate to very high levels, ultimately causing degradation of catalyst performance, with serious effects on process efficiency. A tendency toward higher feed metals levels is a natural consequence of processing generally heavier, more asphaltenic crudes [116] or cutting more deeply into a given crude [28], but, as shown in Fig. 27, may also arise from unit upsets or major variations in operating practice. For instance, the bulk of metals originally present in a crude usually will eventually become concentrated in residua such as vacuum tower bottoms [134], but an upset may cause considerably more than trace amounts of organometallics to become entrained and carry over in the gas oil overhead—a stream destined for FCC feed.

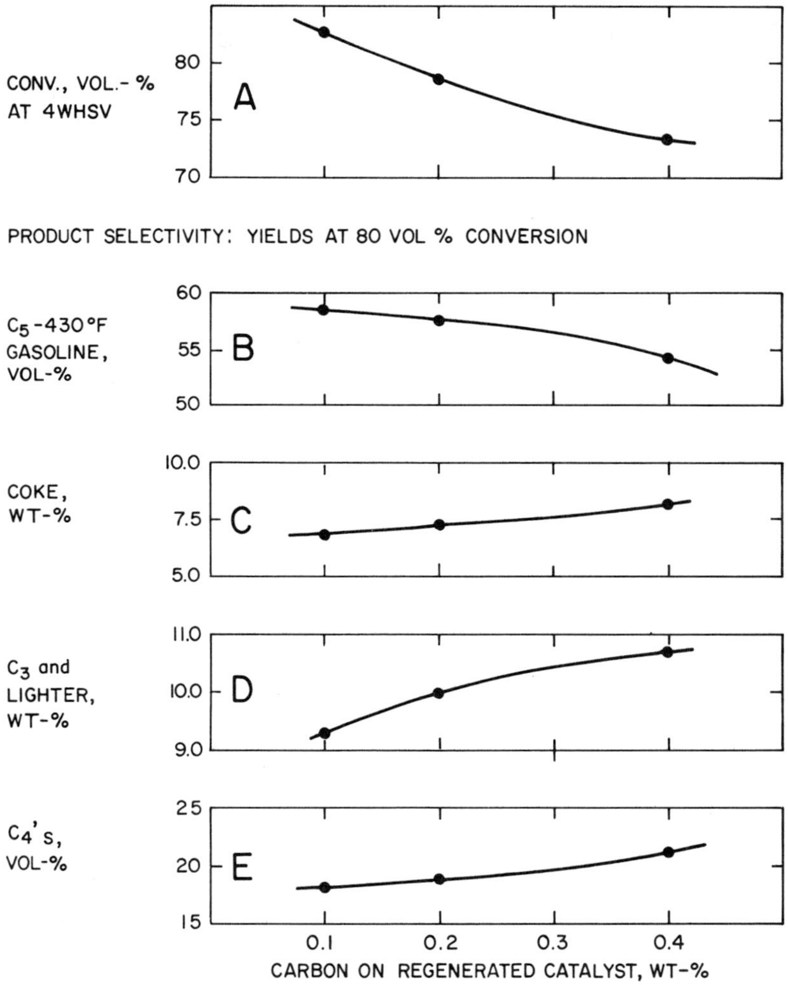

FIG. 26. Conversion and selectivity trends with residual carbon on catalyst for pilot-plant cracking of 27° API, 570° F+ Mid-Continent gas oil over commercial zeolite-containing equilibrium FCC catalyst. Data are adapted from Wachtel et al. [129]. Catalyst contains (ppm) Ni (478), V (1203), and Fe (3730), and shows carbon producing factor (CPF) of 0.75. Based on an analysis by Stuart [259]. Average metals levels (for over 85% of United States and Canadian FCC units) for October-December 1974 were (ppm) Ni (264), V (472), and Fe (3400), with ranges (ppm) of 63 to 2266, 101 to 6805, and 1400 to 8200, respectively.

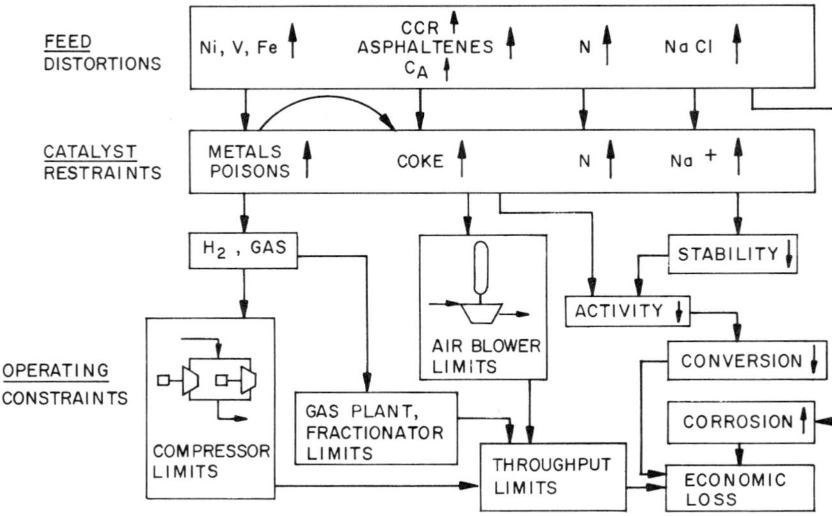

FIG. 27. Wide swings in feed composition and operational upsets are coupled with feedstock distortions which consequently affect catalyst performance and unit operations. Thus, use of heavy crude, clarified slurry oil, or upsets in hydrotreater, vacuum tower, coker, etc., may impose high loads of coke precursors of low H/C ratio and heavy metals contaminants. Desalter failure can introduce high levels of NaCl.

An increase in the level of (active) catalyst metals is generally associated with a moderate decrease in conversion (lower activity) [39, 134, 260] and a deterioration in selectivity—the onset of nonselective cracking [28]. This is evidenced by increases in hydrogen and coke [12, 29, 39, 72, 102, 116, 118, 171, 260, 261], and a decrease in gasoline [28, 29, 39, 72, 260]. Increases in dry gas [28, 29], product olefinicity [12, 116, 118], and CO_2/CO ratio in regenerator flue gas [12, 134] have also been reported. This type of dehydrogenation activity pattern [12, 72, 116] is exemplified by the data of Habib et al. [260] in Fig. 28. Note that the sharp increases in hydrogen and coke with the onset of metals poisoning are nearly parallel, but that further additions of metals above the 500 ppm Ni-equivalent showed smaller incremental effects. In addition to generation of hydrogen, changes in other dry gas components were observed, including a very slight increase in methane and declines in C_3/C_4-olefins and isobutane [260]. The H_2/CH_4 ratio is generally regarded

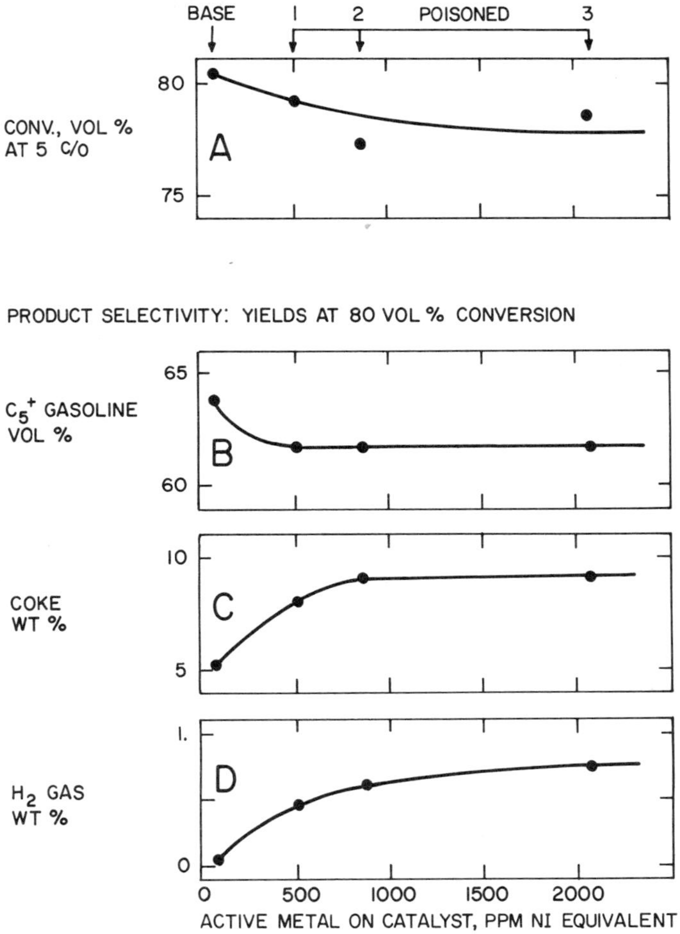

FIG. 28. Conversion/selectivity trends in short contact time riser FCC cracking of 650° F+, 24.6° API hydrotreated Arab Light atmospheric residuum at 1000° F over commercial equilibrium zeolite catalyst artificially poisoned to several metals levels. Data from Habib et al. [260].

as a fairly sensitive indicator of metals-induced dehydrogenation activity [12, 28]. As also observed in the study of Cimbalo, Foster, and Wachtel [28], the maxima in the profiles of gasoline yield vs conversion (where gasoline recracking begins to predominate over gas oil cracking) shifted to lower conversion levels with metals poisoning [260]. It has also been observed [261] that at very high metals levels, actual physical deterioration of the catalyst may cause problems under certain operating conditions.

The increased yields of hydrogen and light gas observed with metals poisoning impose a serious load on refinery gas-handling facilities [12, 28, 134, 260]. This has a chain of consequences that affects the entire system, including riser/reactor, main column, condensers, and compressors, and may force a cutback in unit throughput [102]. The complex coupling resulting from such interactions is visualized in Fig. 27. Large metals-induced molar expansions (mainly hydrogen) such as shown in Fig. 29 are capable of having considerable effects on the performance of wet gas compressors [12, 28, 221, 223]. Such compressors are usually of centrifugal design, and the high hydrogen yields present problems not only because of the large volumes, but also since the low gas density limits the maximum pressure ratio attainable per stage [260].

Because commercial FCC units operate in a heat-balanced mode (see Section VIII-B-1), it is necessary for coke yield to be within a fairly narrow range, say around 6% [50], to be able to meet the energy balance constraint. Thus the increased load of coke-on-catalyst associated with metals contamination will reduce catalyst circulation rate, lowering severity and conversion, and may bottleneck existing regenerator coke burning facilities [28, 39, 50, 102, 134, 260]; i.e., air blower capacity may not be adequate, or there may be regenerator temperature limitations, etc. [17]. The consequences of such a "coke bottleneck" are shown schematically in Fig. 27.

Based on an excellent classification by Cimbalo, Foster, and Wachtel [28], with some modifications by Habib et al. [260], the coke which arises during fluid catalytic cracking may be assigned into four categories:

1. "Catalytic" coke, which is generated in association with acid-catalyzed cracking, increases exponentially with conversion, is lower with zeolite catalysts [29], and is residence time-dependent [166]. Catalytic coke, as a percent of feed, is usually directly proportional to the C/O ratio (other conditions being constant [260]).
2. "Contaminant" coke, which derives from catalytic dehydrogenation action of metals poisons such as nickel, vanadium, etc.
3. "Additive coke" (or carbon), C_{Add}, which is that fraction of the coke which is independent of the C/O ratio or catalyst type, instead correlating with the basic nitrogen and average molecular weight of the feedstock [114] as well as the Conradson Carbon Residue [28, 29].

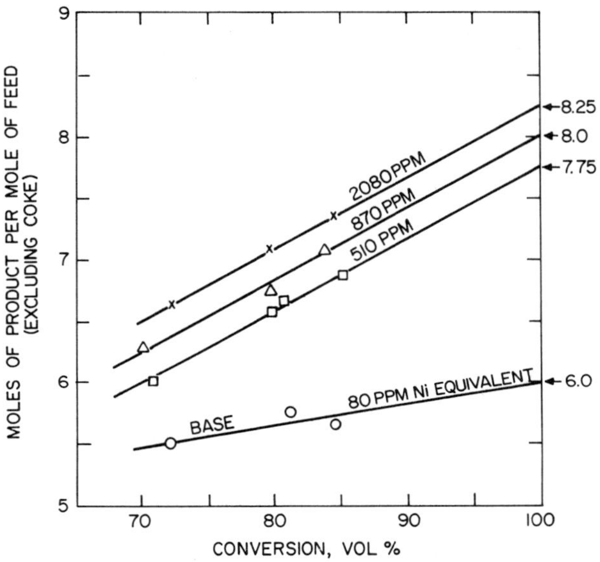

FIG. 29. Plot of total molar expansion vs conversion for short contact time riser cracking of hydrotreated Arab Light atmospheric residuum over equilibrium zeolite catalyst artificially poisoned to several metals levels; data from Habib et al. [260]. Note the very large increase in molar expansion (i.e., the extrapolated 100% conversion values) with increasing metals poisoning; this increase is due almost entirely to the increased yield of molecular hydrogen, which, although small on a weight basis, is very large on a molar basis.

4. "Cat-to-oil" coke, which is a measure of unstripped but potentially strippable hydrocarbons; these are not really coke, but because of nonideal stripping in the commercial process, are also burned in the regenerator.

Occasionally, alkali or alkaline earth metals may appear in FCC charge, usually via leakage of feed water from steam generator tubes on the main fractionator [102] or from faulty crude desalting [51, 102] (see Section III-B-2). Metals such as sodium, lithium, and potassium (or even calcium, barium, etc.) will deactivate FCC catalyst by both "neutralization" of acid sites and reduction of thermal and hydrothermal stability of catalyst [102] (see Section VII-D and Fig. 27).

Hydrotreating has, in the past, been a major means of removing coke or metals restraints where increased yields of gasoline were required from highly aromatic feeds or cycle oils [39, 92, 226]. Similarly, solvent deasphalting (SDA) with propane or even butane and pentane solvents has been utilized to remove coke precursors [116, 137]. A solvent technique similar to SDA has also been used for demetallization [39, 57, 108, 136], wherein a high metals-content residuum is split into a demetallized oil and asphalt. Of course, use of a (zeolite) catalyst with better selectivity for coke [17, 94] and proper operating conditions [39] can substantially aid in removing a coke bottleneck. Attention to catalyst management—increased rates of addition/withdrawal—will also aid in coping with a metals problem [39, 72, 134, 261]. A makeup rate as high as 0.6 to 0.7 lb/bbl was cited for the case of direct cracking of Texas Panhandle resid [39]. "Scavenging"-type demetallization processes have been proposed [134, 136, 261], where a slipstream of catalyst is withdrawn continuously, processed in a demetallization circuit, and returned to the FCC; an earlier version [134] employed ion-exchange resins.

D. Restraints in Gas Plant and Fractionation System

One of the consequences accompanying the use of zeolite cracking catalysts—with their higher first-pass cracking activity [94]—was a larger fraction of gasoline in reactor effluents [74, 77]. As a result, many existing (main column) fractionators became overloaded [76, 94] and sharp separations were no longer possible. This was particularly noticeable in the upper sections of the column [74], with increased loads on the overhead condensors [94]. In the lower section, where there was decreased or eliminated recycle, less heat was available in the main column bottoms reflux for reboiling, feed preheat, or steam generation [74, 94]. Because of the greater liquid (gasoline) yield with zeolites, there was also overload on the absorbers and fractionators in some gas plants [94], notably in the debutanizer, which in some instances required as much as a 10 to 15% increase in capacity [94].

Thus there were many unit revamps where the capacity of fractionating equipment was increased [94], notably more overhead condensing and heat exchanger surface, enlarged gasoline pumps, and a trend toward higher pressures in the main column so smaller towers could be used [77, 94]. This last approach tended toward higher horsepower requirement for the regenerator air blower, which had to pump into a higher pressure regenerator [77, 94].

In some cases the lower ratio of C_2^- gas/gasoline had some benefits in reduced compressor load, the latter being mainly set by yield structure (and the temperature of the available cooling water [74].

Hydrogen blistering and corrosion acceleration by cyanides are frequently encountered in FCC gas recovery systems [262, 263], particularly at high process severity and when sour crudes with high nitrogen content are used [263]. Cyanides remove the partially protective film of iron sulfide on vessel walls, thereby greatly increasing initial corrosion rates [263]. Sometimes, particularly if inadequate wash water is used, Prussian Blue [$Fe_4(Fe(CN)_6)_3$, ferric ferrocyanide] or ferricyanides can form in substantial quantities [262]; these can decompose upon heating (i.e., welding) to give off highly toxic HCN. Hence they must be removed by mechanical means such as sand blasting.

E. Aspects Related to Product and Environmental Restraints

Confronted by the imposing array of impurities that have accumulated in feedstocks from natural processes (Section III-B-2), the refiner is under pressure to evolve technology to minimize or eliminate their effects. He may be restrained by environmental (pollution) regulations, by process performance considerations, or by product specifications. The impact of such constraints is visualized in Fig. 30, and encompasses the spectrum of gaseous emissions, particulates, and liquid effluents and products issuing from the FCC complex.

1. The Problem of Sulfur and Other Impurities

There is great current emphasis on controlling sulfur emissions in the United States [264]. Huling and Readal [264], and Wollaston, Forsythe, and Vasalos [126] found that the distribution of sulfur among the various products of cracking was highly dependent upon the type of sulfur compounds initially present in the feedstock, as well as on process conditions [126]. Huling and Readal [264] found wide variations, with ranges of 19 to 60% for H_2S, 3 to 10% in gasoline, 11 to 35% in light fuel oil, 7 to 34% in decant oil, and 3 to 28% in coke. Table 9 illustrates these variations—and the actual sulfur levels in the major fractions—for a number of feeds at about the same (78 to 80 vol%) conversion level.

The desulfurization that accompanies cracking "generally" removes about 40 to 50% sulfur as H_2S at typical conversion levels [29, 39, 126], with much of this deriving from the cracking of C—S bonds in nonthiophenic compounds by catalyst acid sites [126]. Feeds that

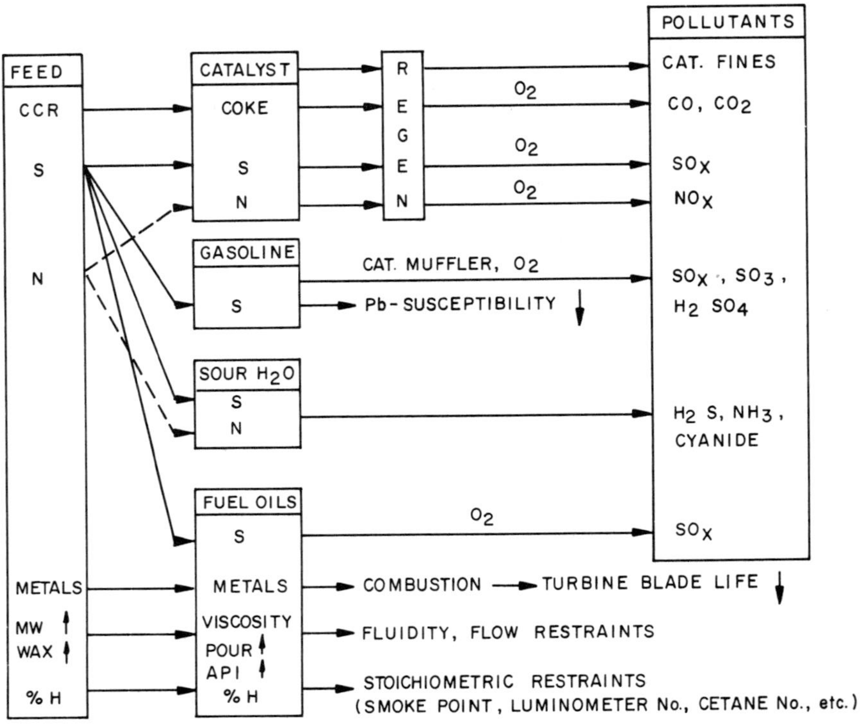

FIG. 30. Naturally occurring impurities in FCC feedstocks can lead to difficulties in meeting product specifications, particularly federal, state, and local environmental restrictions.

desulfurize most readily when cracked (i.e., a large fraction of feed sulfur goes to H_2S) generally show the highest fraction of feed sulfur in gasoline and the lowest fraction in coke [264] (see Table 9). Note (Fig. 31) that sulfur distribution is highly dependent on conversion [126]. As much as 3.8% of feed sulfur converting to mercaptans has been reported from the cracking of virgin gas oils [39]; about 1.9% of feed sulfur in the (overhead accumulator) sour water has also been observed [39]. Such sour water also may contain small amounts of ammonia, phenols, and cyanides [262], and should not be used as spray for regenerator plenum and stack gas since they constitute toxic emissions [262]. H_2S, either from FCC gas plants or from light or heavy cycle oil via hydrotreater effluents, can be handled conveniently in amine scrubbers followed by the Claus process and sulfur recovery [29].

TABLE 9

Sulfur Distribution in Products from Short Contact Time, High Temperature Riser FCC Cracking of Several Different Feedstocks[a, b]

	West Texas virgin gas oil	HDT West Texas gas oil	California gas oil	Kuwait virgin gas oil	HDT Kuwait gas oil	Kuwait[c] DAO/gas oil blend
Feed, wt% S	1.75	0.21	1.17	2.66	0.11	3.14
Conversion, vol%	77.8	77.8	78.7	78.6	80.4	80.1
S distribution, wt%[d] of Feed Sulfur						
H_2S	42.9	19.2	60.2	46.5	29.0	50.0
Light gasoline	0.2	0.9	1.6	0.3	0.1	1.9
Heavy gasoline	3.3	1.9	7.9	3.5	2.2	5.0
Light fuel oil	28.0	34.6	20.7	21.1	36.9	17.3
Decanted oil	20.5	34.7	6.8	17.3	19.9	15.3
Coke	5.1	8.7	2.8	11.3	11.9	10.5
S content, wt%						
Light (C_2 to 250° F) gasoline	0.02	0.01	0.07	0.04	0.00	0.17
Heavy (250 to 430° F) gasoline	0.25	0.02	0.37	0.54	0.01	1.06
Light (430 to 650° F) fuel oil	2.67	0.53	1.48	5.06	0.26	4.19
Decanted oil (650° F+)	4.20	1.35	1.68	6.86	0.88	6.06
Coke	2.23	0.49	0.52	4.42	0.28	4.15

[a] From data of Huling and Readal [264].
[b] High activity zeolite catalyst, 960 to 1030° F, <5 sec residence time.
[c] DAO = deasphalted oil.
[d] No loss basis.

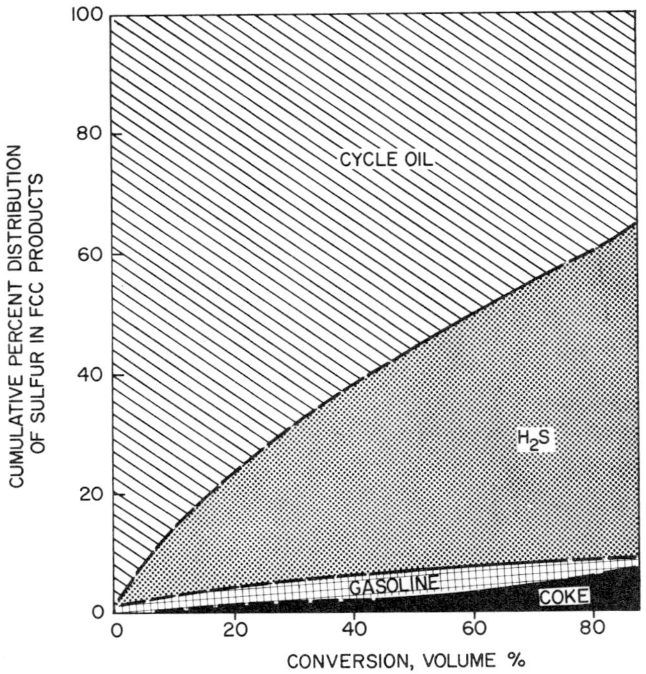

FIG. 31. Plot of sulfur distribution in FCC products vs conversion for a virgin gas oil. Adapted from Wollaston, Forsythe, and Vasalos [126]. Note that sulfur distribution varies with conversions. At "typical" conversions of 70 to 80 vol%, about half the sulfur is converted to H_2S, while most of the rest remains in cycle oil, with small amounts distributed to gasoline and coke-on-catalyst; this contrasts with thermal cracking where H_2S yield is 10 to 20%.

Thiophenic sulfur compounds crack much more slowly, and uncracked thiophenes with one, two, and three or more rings can be found in gasoline, light cycle oil, and heavy cycle oil/clarified slurry oil, respectively [126]. Sulfur in any product from catalytic cracking is undesirable, and is particularly troublesome in gasoline because it reduces the susceptibility to lead compounds for enhancing octane number [102]. Also, combustion of sulfur compounds (i.e., thiophene) in gasoline results in formation of sulfur dioxide in automobile exhaust, and further oxidation to sulfur trioxide and sulfuric acid over catalytic mufflers has been reported [265]. It has been observed that the level of sulfur in clarified slurry oil is generally proportional to that in coke [264]. The fraction of FCC feed sulfur ending up in coke is related to both sulfur type and Conradson carbon in feed as well as coke yield [264].

Of course, sulfur that is situated in coke (on catalyst) is highly undesirable because of the subsequent emission of sulfur oxides (SO_x) in the regenerator flue gas [47, 72, 102, 126, 226]. In fact, total stack gas emissions from an FCC—including SO_x, nitrogen oxides, carbon monoxide, and particulates—constitute a major environmental pollution concern [226], potentially raising the need for expensive stack gas desulfurization (scrubbing) [72, 92, 120, 126, 226] or feed hydrotreating [59, 65, 92, 126, 147, 226, 264] in order to meet air pollution regulations. In addition, techniques to remove carbon monoxide—such as CO-boilers [94, 226] and CO-combustion catalysts [240]—may be required. Figure 32 shows a correlation [264] between the expected SO_x content of FCC regenerator CO-boiler stack gas and the sulfur content of FCC feedstock. The Texas Air Control Board limit for existing equipment is 400 vppm SO_x emissions from the CO-boiler stack [264]. Note from Fig. 32 that various stocks must be desulfurized to different levels to meet this limit.

2. Effect of Hydrotreating

As discussed in Section VI-B-2, hydrotreating reduces the sulfur concentration of all FCC products, especially in the gasoline and light fuel oil [47, 147]. The fraction of feed sulfur in heavy cycle oil or clarified slurry oil, however, may actually increase [47, 130, 147] in hydrotreated (HDT) stocks due to the preferential removal of light sulfur compounds, with retention of higher molecular weight multiring thiophene compounds which are less reactive [47, 130, 147, 264]. However, HDT appears to reduce sulfur in FCC coke somewhat more (percentage-wise) than for the feed as a whole [126, 147, 264], and can sharply lower the expected SO_x emissions from FCC stack gas effluents [59, 118, 147, 264]. McCulloch [147] observes that for a 50% vacuum gas oil-50% FCC cycle oil feed (23.6 °API, 1.32% S, 36% aromatics), HDT can reduce gasoline sulfur to 30 ppm, well below the 100 ppm maximum level.

F. The Regenerator Zone: Some Restraints and Interactions

1. Intrinsic vs Diffusion—Limited Coke Burning in Bead Catalyst

Before discussing regeneration from the standpoint of the FCC process, it is worthwhile examining the dynamics of coke burning

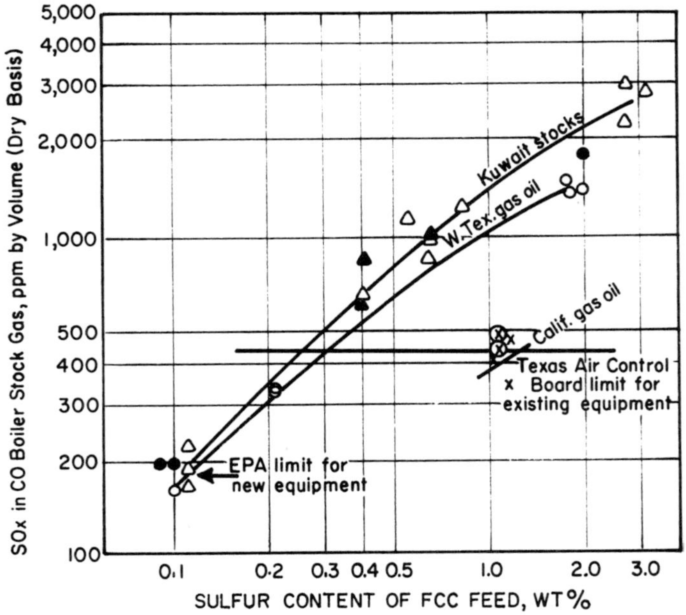

FIG. 32. Correlation between SO_X content of CO-boiler stack gas vs % sulfur in FCC feed as adapted from work of Huling, McKinney, and Readal [264]. SO_X contents are calculated from sulfur in coke (as derived from flue gas sulfur analysis) and related to commercial regenerator operation by assuming an air/coke ratio of 11/1 and a CO_2/CO ratio of 1.3, that FCC regenerator flue gas is the only fuel to the CO-boiler, and that the boiler is operated to leave 2.0% O_2 in the stack gas. CO-boilers commonly reduce CO-level of effluent flue gas to about 0.05% before release to the atmosphere.

in TCC bead catalysts. The classic work of Weisz and associates [168, 218, 220, 266, 267] not only provides a strong kinetic frame of reference, but also demonstrates a subtle interplay of restraints. By "coke" we mean the carbonaceous, combustible residue remaining on catalysts after cracking, such deposits showing stoichiometry in the range of $C_{1.0}$ $H_{1.0}$ to $C_{1.0}$ $H_{1.5}$ [266].

With porous solid catalysts such as silica-alumina, there exists an intrinsic, uncatalyzed burning rate that qualitatively and quantitatively resembles combustion of pure graphite [266]. This intrinsic rate, however, is accessible to observation only if perturbing influences such as mass and heat transfer effects, and unstripped hydrocarbons, have been avoided, say with thoroughly stripped bead catalyst that is used in a thin bed after sufficient grinding to pass through a 200-mesh screen [266]. The intrinsic rate of coke removed is first-order in carbon, i.e., d[c]/dt = -k[c], where k is a first-order rate constant and is independent of both the initial amount of coke and the fraction of burnoff. The rate of carbon burnoff for a given support appears to be largely independent of the source and structure of coke [266].

Weisz and Goodwin [266] also noted that the presence of transition metal oxides resulted in coke burning rates up to several orders of magnitude higher than the uncatalyzed rates (at a given temperature). Thus when 0.15 wt% Cr_2O_3 was added to the bead, the burning rate was first-order, but k was four times greater than in the absence of Cr_2O_3; however, the catalytic activity was lost if the catalyst was treated with high temperature steam or after numerous cycles of cracking and regeneration.

There also exists an intrinsic ratio of CO_2 and CO—the products of coke combustion—in spite of the fact that a broad spectrum of CO_2/CO product ratios has been reported in the literature for combustion of carbonaceous deposits on porous catalysts. Thus Weisz [267] has shown that there is a definite initial ratio of CO_2/CO produced at the burning site which is dependent on temperature only, and which ratio agrees exactly with that of Arthur [268] for graphite burning at the same temperature. This initial ratio, however, is increased by subsequent intraparticle conversion of CO to CO_2, which is in turn dependent on the temperature, the catalytic rate constant for CO-oxidation, and the rate of diffusional escape of CO from the pore-channel system of the catalyst [267]. Thus larger particle size (longer diffusion path) and lower diffusivity (more resistance to movement within the pores—as well as the presence of Cr_2O_3 as catalyst) would tend to increase secondary conversion (and therefore the apparent CO_2/CO ratio) [267].

Weisz and associates [218, 267] further showed that in typical silica-alumina bead cracking catalysts, the coke combustion rate as a whole begins to show significant inhibition by oxygen diffusion at temperatures above about 500° C (932° F), leading to local coke-removal rates that decrease toward the center of the bead. This causes progressive lengthening of the diffusion path during oxidation, producing great variance in CO_2/CO ratios. Stated differently, at 500 to 600° C (932 to 1112° F), coke burning shifts from the intrinsic region where coke burning is radially uniform within the bead, to a region where shell-progressive, diffusion-controlled combustion is dominant [218] (see Fig. 33).

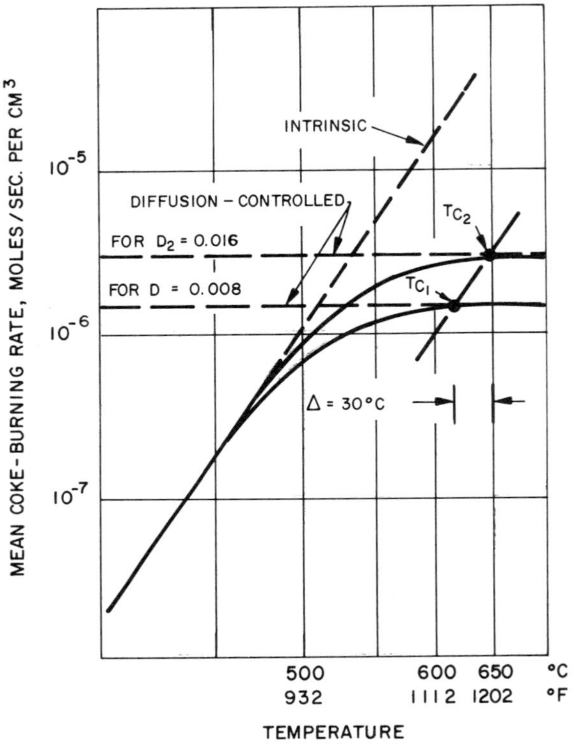

FIG. 33. Plot of mean coke burning rate vs temperature; after Weisz [220]. The actual rate of coke burning vs temperature lies between the lines corresponding to intrinsic and diffusion-controlled combustion. The intrinsic burning line shifts up or down in relation to initial coke concentration and pO_2, while diffusion-controlled lines shift with internal catalyst diffusivity (D) and pO_2. A critical temperature (T_c) associated with a particular degree of boundary sharpness of the coke shell delineates the transition between intrinsic and diffusion-limited kinetics.

Weisz and Goodwin [218] developed the following equation to describe the kinetics operative in the shell-progressive region for bead catalysts:

$$\tau_{85} = 0.076 \frac{R_0^{\ 2} C_c}{nDC_{ox}}$$

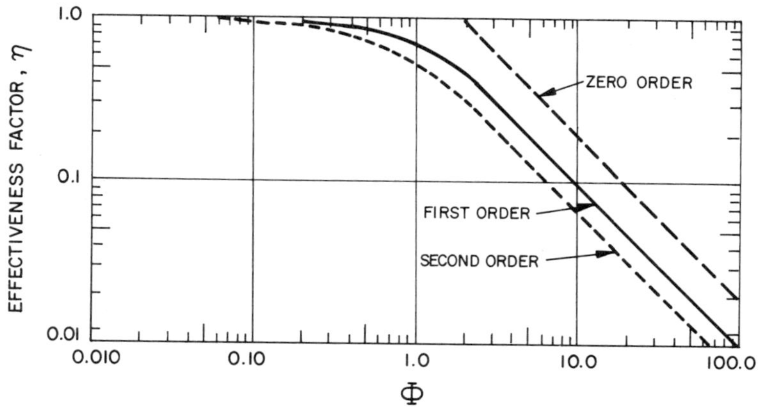

FIG. 34. The Φ criteria as exemplified for zero-, first-, and second-order reactions. After Weisz [168].

where τ_{85} = time required for 85% coke combustion at a given temperature

R_0 = radius of particle

C_c = initial molar carbon concentration on solid catalyst

n = stoichiometric ratio of carbon burned/molecule of O_2 consumed

D = effective diffusivity of porous catalyst for O_2

C_{ox} = oxygen concentration

Thus, in the diffusion-controlled region, the time of burnoff is dependent upon initial carbon content, particle size, diffusivity, and oxygen partial pressure, but is independent of catalyst composition and nature of the coke [218]. Note also that there would be no kinetic consequences of positive internal temperature gradients due to exothermic reaction processes [218].

In demonstrating where parameters such as size and diffusivity begin to have important influences on catalytic processes, Weisz and associates [168] related the Thiele effectiveness factor to a parameter Φ (see Fig. 34) which correlated with experimentally observable variables:

$$\Phi = \frac{dn}{dt}\frac{1}{c}\frac{R^2}{D} \geqq \sim 1$$

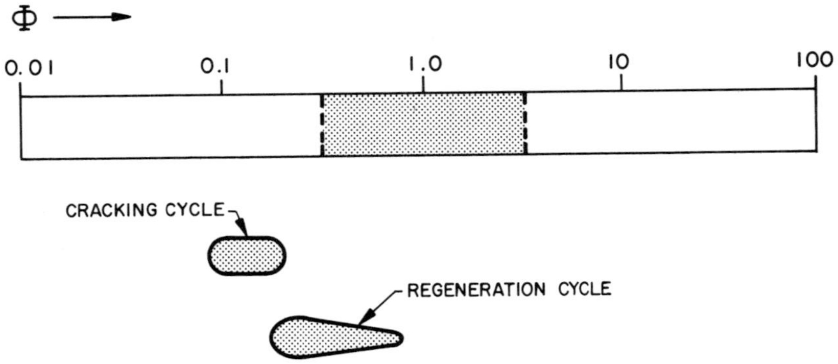

FIG. 35. The effect of diffusion restraints on the TCC process; after Weisz [168]. Diffusion limitations occur when Φ is greater than the region of unity (dotted area). Although the Φ-criteria shows no diffusion problems in the cracking cycle, coke combustion, particularly if temperatures are high, can sometimes enter a diffusion-limited regime in regeneration of TCC bead catalysts.

where dn/dt = actual observed reaction rate/unit volume of catalyst particle
C = externally-applied reactant concentration
R = size (approximate equivalent particle radius)
D = diffusivity of reactant (in the particle medium)

Weisz showed that, if Φ approaches or is greater than the region of unity, then significant diffusional influences will be experienced [168]. The zeolite catalysts, with activities—and hence dn/dt—2×10^3 to 10×10^3 times higher than the older amorphous varieties, might have been expected to encounter diffusion resistance in TCC beads of 0.3 to 0.4 cm diameter [168] (see Fig. 35). Here the use of FCC technology was important in taking full advantage of high zeolite activity, since FCC particles have a size (R) many times smaller than bead catalysts [168].

Weisz and associates [168, 220] also noted that not only did sharp radial gradients in residual carbon concentration appear when coke burning rates become diffusion-inhibited, but in some cases a finite number of beads remained with sharp coke cores. This is the so-called "fish-eyed" bead [220]. This situation predisposed the bead to internal strains and a tendency toward mechanical fracturing.

Thus, when subjected to repeated cycles of coke deposition and regeneration in commercial practice, a high rate of catalyst attrition through breakage was observed. When, however, a binodal pore system was incorporated into the catalyst particle [167], overall accessibility and diffusivity were greatly improved. The calculated Φ value no longer fell into the diffusion-controlled region, and attrition losses declined [168, 220]. The magnitude of this type of effect can be seen in Fig. 33. Doubling diffusivity D (from 0.008 to 0.016 cm^2/sec) leads to an increase of 30° C in the temperature, T_c, below which there is radially uniform burning. This allows removal of significantly larger amounts of coke before onset of diffusion inhibition.

2. Problems and Interactions in the FCC Regenerator

The cracking reaction rate—and hence coke formation rate—cannot increase beyond the corresponding coke oxidation rate in the regenerator if a steady-state operation is expected [101]. Further, it is in the harsh environment of the regeneration zone that most of the stress on the FCC catalyst occurs. It is an objective of regeneration to obtain a low level of carbon on regenerated catalyst [20, 65, 77], at an efficient and sufficiently high burning rate to reduce catalyst residence time [77, 226], thereby minimizing exposure of catalyst to high-temperature deactivation—and activity loss [269]. Some of the important factors governing efficiency of regeneration [11, 74, 118] include (1) initial level of coke on catalyst, (2) temperature, (3) oxygen partial pressure, (4) efficiency of air distribution and air/catalyst contact, (5) inventory (average residence time), and (6) catalyst type (composition, activity, impurities, etc.).

Factors Related to Catalyst Activity and Inventory

Catalyst activity decay derives from two different phenomena [270]: (1) the short-term aging from coke deposition which is counteracted by regeneration, and (2) a long-term aging which is due primarily to permanent damage to some of the active sites on the catalyst surface. Thus fresh (makeup) catalyst must be periodically added—and "aged" catalyst withdrawn—to maintain a desired level of average activity [65] in the unit. Activity can be correlated with makeup expressed as percentage of inventory/day [65].

Thus it is desirable to have a small catalyst inventory [62, 65, 66, 74, 94, 271], the minimum level consistent with regeneration kinetics [65] (i.e., what residence time is required to burn the carbon on spent catalyst down to about 0.1 to 0.2% on regenerated catalyst). A smaller inventory allows a faster unit response to catalyst makeup [62] and a more economic makeup policy [65, 66] (less costly fresh catalyst required to maintain a given equilibrium activity). It is also associated with less attrition [74]. Controlled withdrawal of FCC catalyst also allows removal of portions of inventory that have been metal-contaminated [29] or have deteriorated into inert, high-density microspheres with a tendency toward thermal cracking and coke formation [76].

Lee [270] has formulated a method for obtaining optimal policy for catalyst makeup. The rate of deactivation (loss of activity A with time t) in a perfectly mixed system generally follows the relationship

$$dA/dt = -kAt^n$$

where k and n vary with the catalyst, and k is a function of temperature [270]. Makeup rate S equals withdrawal rate plus losses through cyclones (to regenerator stack and—through reactor effluent—to slurry oil). At long times the steady-state (equilibrium) activity can be described as

$$A_s = \frac{A_a S}{S + k}$$

where A_s and A_a are steady-state and fresh catalyst activities, respectively. Thus if one keeps on adding catalyst makeup at rate S, with fresh activity A_a and deactivation rate constant k, the average activity will eventually equilibrate at a steady-state activity A_s. But as A_s (and thus cracking rate) increases, there may be penalties on selectivity such as overcracking to gas and coke, requiring judgment on how high a level is really optimum. Feedstock and product slate must also be considered in assessing the optimal makeup rate.

Attrition, Catalyst Losses, and Cyclone Performance

For smooth regenerator operation, it is important to have good catalyst flow patterns [269, 271], including good aeration of catalyst lines [74] and proper design of catalyst inlets, protecting baffles, and outlet configuration [269]. High vapor velocities through the dense bed improve diffusion [65, 271] and fluidization [269] characteristics. However, excess air velocities can lead to catalyst attrition and losses [62, 65, 77, 81], or to erosion of linings [81, 269], grid holes, slide valves, etc. [63], and hence to shorter service life [63, 73]. In this regard, Bunn et al. [62] report that less catalyst losses (via entrainment) occur if the upper section of the regenerator has a larger diameter (and thus permits lower linear velocities).

One of the earliest attempts to prevent solids emissions featured the use of spray-dried microspheroidal catalyst. Later, improved cyclones and cyclone operation [62, 84, 269], particularly two-staged cyclones [77], were employed, largely for economic reasons. The makeup of catalyst losses was very expensive. Recently, however, the thrust has been concern for atmospheric emissions and the imposition of stricter local laws and law enforcement, and considerable effort has been directed toward preparation of attrition-resistant catalysts [50, 84, 92, 226, 272]. Some electrostatic precipitators (Cottrell) have also been installed [94]. Catalyst attrition resistance or hardness (i.e., resistance to formation of fines less than 10 μm in diameter) has been cited [84] as more important than particle density in minimizing catalyst entrainment to regenerator cyclones.

Erosion in cyclones can be minimized through erosion-resistant refractory linings [73]. Cyclone and dipleg support systems are carefully designed to minimize potential damage from stress and vibration [64]. A satisfactory pressure balance between the cyclone system and the regenerator vessel can avoid large pressure differentials that could prevent flow of solids downward into the regenerator—thus lowering the risk of blowing the dipleg seals [273]. Improved cyclone configuration allows the maintenance of a high fines level in the circulating catalyst, thereby improving its fluidization properties [73]. If power recovery is desired, a third-stage cyclone (separator) is required [273], since a low (solid) fines content is important for long turbine blade life [65].

Catalyst/Air Mixing and Flow Patterns

Good distribution of combustion air and efficient catalyst/air contact are critical in the regeneration zone [94, 118, 240, 269, 271]. This is necessary to sustain an even rate of coke-burning [76], to insure uniform fluidization [269], and to avoid actually bypassing certain regions of the regenerator vessel [74]. Improved mixing will help to decrease tendencies toward localized high temperature deactivation of catalyst [62, 77, 78, 269], i.e., as can occur when uneven catalyst/air flow patterns lead to formation of large bubbles with high local concentrations of oxygen [269]. Improved mixing is usually obtained with air-grids [92, 226, 269], by modifying holes in grids or nozzle design [94, 240], by increasing the regenerator height/diameter ratio [92, 226], etc. Maximum utilization of air is necessary to obtain a low carbon level on regenerated catalyst and with minimum afterburning [65]. It has been suggested that, ideally, catalyst/air contact should approach a perfect mixed condition, where carbon levels on catalyst in the bed should be about the same as those on the catalyst leaving [74]. Good distribution of oxygen is particularly important when regenerating in the CO-burning mode [39, 240].

Effect of Higher Regenerator Pressure

Higher regenerator pressures (i.e., up to 36 psig [269]) have been cited as advantageous because of consequent higher oxygen partial pressures—and improved regeneration kinetics [65, 92, 240, 271] and more economical power recovery [77, 88, 89, 116, 118]. Note that utility requirements for compressing air can be covered by power recovery utilizing the ΔP between the regenerator operating pressure and the atmosphere via a flue gas expander [116]. Hot gas power recovery in the FCC does, however, require higher investment. After removal of fines in the third-stage separator (to prevent severe erosion of turbine blades by catalyst fines), flue gas passes to expander turbines which, once start up is accomplished, supply more than enough power to drive the air-blower [88, 89]. Conventionally, the flue gas is expanded through the slide valve and orifice baffle chambers.

Regenerator Temperature and its Implications

There has been a general evolution toward higher regenerator bed temperatures [62, 65, 101, 226, 269]; average dense bed temperatures (T_{RG}) have increased from about 1150° F in the 1950s, through about 1200° F in the 1960s, to the range 1250 to 1300° F in the 1970s [20, 92, 226, 240, 271]. This trend derives from the higher coke-burning rates and burning efficiency afforded by higher T_{RG} [74, 245], the drive toward lower carbon levels on regenerated catalyst to promote zeolite accessibility [65], from the trend to convert more CO to CO_2 in the regenerator [94], and as a result of the trend to higher reactor temperatures. In order to effect higher temperatures without encountering a metallurgical restraint, there has been a shift from carbon steel or low alloy internals to high alloy (i.e., high nickel-chromium stainless) steels [20, 65, 78, 92, 94, 150, 226, 240], particularly for cyclones, linings, hangers, stack slide valves, etc. Conversely, a fixed regenerator temperature limitation may lower potential conversion and gasoline yield [272].

On the other hand, considerable care must be taken to obtain good air distribution and catalyst flow patterns to avoid local hot spots and resultant catalyst deactivation by sintering [62, 74, 78, 94, 269].

Of particular concern are the transient high temperatures (as high as 1600° F) that can occur when catalyst microspheres coated with reactive oil or unstripped heavy hydrocarbons are exposed to high local partial pressures of oxygen [76, 269]. In this regard, use of torch oil or clarified slurry oil to help increase T_{RG} may have deleterious effects [94]. Note that the use of feed preheat (FPH) and high reactor temperatures aid in increasing T_{RG} [20, 74, 94] (see Section VIII-B-2). It also has been reported that higher T_{RG} are not associated with increased formation of nitrogen oxides in flue gas [249].

Oxygen Partial Pressure, Air Requirements, and CO_2/CO Ratio

The carbon burning rate in the regenerator increases with increasing oxygen partial pressure [11, 65, 74, 94, 118]. If one assumes that the regeneration gas flows through the bed in plug flow, the effective oxygen partial pressure would become the log mean of values at the bottom and top of the bed [74]. With good air distribution and efficient regenerator operation, oxygen levels in the flue gas should be less than 1%, preferably 0 to 0.2% [12]. Ideally, the objective is to obtain a low level of carbon on regenerated catalyst while maintaining an oxygen level in the flue gas low enough to avoid problems with afterburning of CO to CO_2 [62]. If excess oxygen does appear in the flue gas, it may be due to a low coke-burning rate [245].

With satisfactory distribution of air (and thus only trace amounts of oxygen in the flue gas), one consequence is that the number of pounds of air required to burn 1 lb of coke becomes essentially constant [12]. This "air/coke ratio" varies between about 11 and 14, depending on the specific mode of regeneration employed [88]. A value of 11.5 ± 0.2 lb air/lb coke burned was cited by Pohlenz for regeneration of amorphous fluid catalysts [12]; values as high as 14.5 lb air/lb coke have been reported for high temperature regeneration (with complete CO-combustion). With complete CO-combustion, only small increases in regenerator air requirements are, in fact, required [39], since coke-make is also cut significantly [78]. Under any circumstances the aim is to have the desired extent of combustion with the minimum quantity of air [269]. Of course, if, as discussed earlier, there is inadequate stripping and a large quantity of hydrogen-rich product is associated with the coke, air demand—and stress on air blower capacity—will increase.

The ratio of CO_2/CO in the regenerator flue gas is dependent on the mode of regeneration and upon the type of catalyst and its degree of contamination with metals [245]. In many regenerators run at 1100 to 1200° F, CO_2/CO mole ratios have typically been close to one (i.e., ~10% CO_2, 10% CO) [78, 118]. That is, roughly half of the carbon burned was oxidized to CO and the other half all the way to CO_2. Naturally, the CO_2/CO ratio has a marked influence on the catalyst circulation rate (due to the increased heat of combustion of coke with conversion of CO to CO_2) [69]. It is necessary to have a reasonably accurate prediction of the CO_2/CO ratio in order to have the proper relative sizing of the air-blower (and the CO-boiler and feed preheat if required) [64].

Heavy metals contaminants (i.e., Fe, V, Ni, Cu) have been shown to promote uncontrolled CO-burning (afterburning) in the FCC regenerator, and units where the catalyst inventory has a high level of metals almost invariably have high CO_2/CO levels in the flue gas [12, 20, 155]. Particularly conspicuous in this regard is "tramp iron" [102, 155]. Not only is iron a dehydrogenation catalyst, but both combined and "tramp" iron catalyze CO-oxidation to CO_2 in the regenerator [102]. The magnitude of this effect can be significant: CO_2/CO ratios of 1.6, 1.8, and 1.9 to 2.7 with "tramp" iron levels of 0.92, 1.2, and 1.27 wt%, respectively, have been reported [155], together with increased temperatures from afterburning in both dense and dilute phases of the regenerator.

Control of Afterburning and Methods for Complete Combustion of Carbon Monoxide

Afterburning—the uncontrolled combustion of CO to CO_2 that can occur in the dilute phase of the regenerator if excess O_2 is present—can cause serious operating problems [12, 74, 141, 240], notably severe physical damage to catalyst and cyclones [64, 73, 224, 245], particularly to cyclones of carbon steel [20]. Runaway temperatures exceeding 1400° F have been reported [118]. There can also be cycling of regenerator temperatures [12]. As described in the preceding subsection, the afterburning problem is compounded by the presence of metals poisons and "tramp" iron. Generally, as the T_{RG} approaches 1300° F, the likelihood of igniting uncontrolled afterburning in the lean phase is increased [20].

One approach to counter this problem is to use high-strength, high-temperature alloy steels to permit higher cyclone temperatures, both in normal usage and as insurance in the event of afterburning [63]. In older units the use of steam or spray water to lower temperatures was employed [12, 141], but this is undesirable because of potential damage to the catalyst from sintering, breakage, thermal shock, etc. [12]. More ideally, afterburning is avoided by skillful control and trim of combustion air [141]. The higher velocities used to improve air/catalyst contact aid in minimizing the effect of uncontrolled afterburning on the catalyst [20]. This has been ascribed [20] to the presence of more catalyst as a potential heat sink in the upper (afterburning) regions of newer regenerators.

In earlier TCC operations, addition of chromium oxide, Cr_2O_3, was practiced, with the objective of catalytically increasing CO_2/CO ratio and preventing uncontrolled afterburning problems [203]. This approach, however, was limited by practical difficulties in regeneration [266] and because the transition metal also functioned as a dehydrogenation catalyst, harming selectivity by forming coke and unsaturates [203]. The most striking and successful method for controlling afterburning is through the use of controlled combustion catalysts [81, 240]. These show high catalytic activity for total CO-oxidation in the regenerator dense phase, even at temperatures as low as 1150° F [20, 240]. A further benefit from this approach is that CO-stack emissions are essentially eliminated or greatly reduced, and a much higher fraction of the total energy released from the combustion of coke is used directly to fulfill the process energy requirements [20, 240]. By rapid burning of CO at relatively low temperatures in the dense bed (where there is a big catalyst heat sink), there is less burning in the dilute phase in the vicinity of the cyclones [240].

Air pollution authorities have recently tightened standards for CO-emissions to the atmosphere. Values of 500 ppm CO for a new source and 500 to 2000 ppm (depending on the state) for an existing source have been cited [78]. A CO-boiler can be used to convert CO to CO_2 [73, 88, 116, 139]. Such a process scheme allows both reduction of CO to permissible levels and higher overall thermal efficiency, wherein the heat of CO-combustion and the sensible heat are recovered in the form of high-pressure steam [88]. However, investment and maintenance costs are high [116]. Before final discharge to the stack, the flue gas may be directed to an electrostatic precipitator or other particulate control device for final recovery of catalyst fines [88]. Improved regenerator designs featuring high temperature operation have been reported as capable of effecting near-complete combustion of CO in the dense bed, thereby minimizing afterburning, and eliminating the need for a CO boiler [20, 66, 78, 79, 116, 118, 240, 249].

IX. CHEMISTRY AND MECHANISM OF CATALYTIC CRACKING

Mechanistic aspects of earlier work have been thoroughly reviewed by Voge [11] and Haensel [274]. We will now attempt to integrate the key aspects of the earlier work with the growing body of knowledge of mechanism over zeolite catalysts.

A. Thermodynamic Aspects

Excellent discussions of thermodynamic factors in cracking have been published by Voge [11] and Wang [241]. Some thermodynamic information relevant to cracking is assembled in Table 10. Although thermodynamic data are commonly of great value in predicting the possible extent of a reaction and the most suitable reaction conditions, they are less useful in catalytic cracking since the key cracking reaction, β-scission, is not equilibrium limited. Thermodynamics can give some limited predictions on equilibria of certain secondary reactions. However, many of the possible hydrocarbon reactions accompanying cracking do not attain equilibrium. Actually, extensive equilibration in a catalytic cracking reaction would result in almost complete degradation to graphite and hydrogen, since that reaction has a favorable free energy change for all hydrocarbons except methane at temperatures above 400° F [11, 245].

As observed by Voge [11] (Table 10), cracking of relatively long chain paraffins and olefins can go essentially (over 95%) to completion at 750 to 1022° F, as can certain hydrogen transfer reactions such as the reaction of a hydroaromatic and olefin to form an aromatic and paraffin. Another group of reactions—olefin double bond and skeletal isomerization, paraffin isomerization, alkylaromatic isomerization, transalkylation and dealkylation, paraffin dehydrogenation, as well as some other hydrogen transfer reactions—are commonly intermediate in attainment of equilibrium. For such systems, observed deviations from equilibrium value may have mechanistic (kinetic) significance; specific calculations must be made (and are rather complex), since isomer ratios will depend upon temperature, pressure, type and quantity of diluents present, and possible simultaneous coupling with other equilibrium reactions involving one or several of the molecular species of interest. On the other hand, condensation-type reactions such as olefin polymerization and paraffin alkylation become disfavored with respect to cracking at high temperatures because of thermodynamic reasons [276].

Heats of reaction may also be calculated from enthalpy data (see the examples in Table 10); the magnitude of these is highly dependent on the feedstock, reaction conditions, and the extent of conversion [11, 241]. From Table 10 it is evident that both endothermic and exothermic reactions contribute to the overall heat of reaction in a catalytic cracking process. Ordinarily, highly endothermic reactions such as cracking and dealkylation predominate at low to intermediate conversion levels [277]. At higher conversions, however, exothermic reactions such as hydrogen transfer, cyclization, and condensation provide increasingly large contributions, and the net heat of reaction

TABLE 10

Some Thermodynamic Data for Idealized Reactions of Importance in Catalytic Cracking[a]

Reaction class	Specific reaction	Log K_E (equilibrium constant) 850° F	950° F	980° F	Heat of reaction, Btu/mole, 950° F
Cracking	$n\text{-}C_{10}H_{22} \rightarrow n\text{-}C_7H_{16} + C_3H_6$	2.04	2.46	-	32,050
	$1\text{-}C_8H_{16} \rightarrow 2C_4H_8$	1.68	2.10	2.23	33,663
Hydrogen transfer	$4C_6H_{12} \rightarrow 3C_6H_{14} + C_6H_6$	12.44	11.09	-	-109,681
	$cyclo\text{-}C_6H_{12} + 3\ 1\text{-}C_5H_{10} \rightarrow 3n\text{-}C_5H_{12} + C_6H_6$	11.22	10.35	-	-73,249
Isomerization	$1\text{-}C_4H_8 \rightarrow trans\text{-}2\text{-}C_4H_8$	0.32	0.25	0.09[b]	-4,874
	$n\text{-}C_6H_{10} \rightarrow iso\text{-}C_4H_{10}$	-0.20	-0.23	-0.36[b]	-3,420
	$o\text{-}C_6H_4(CH_3)_2 \rightarrow m\text{-}C_6H_4(CH_3)_2$	0.33	0.30	-	-1,310
	$cyclo\text{-}C_6H_{12} \rightarrow CH_3\text{-}cyclo\text{-}C_5H_9$	1.00	1.09	1.10[b]	6,264
Transalkylation	$C_6H_6 + m\text{-}C_6H_4(CH_3)_2 \rightarrow 2C_6H_5CH_3$	0.65	0.65	0.65[b]	-221
Cyclization	$1\text{-}C_7H_{14} \rightarrow CH_3\text{-}cyclo\text{-}C_6H_{11}$	2.11	1.54	-	-37,980
Dealkylation	$iso\text{-}C_3H_7\text{-}C_6H_5 \rightarrow C_6H_6 + C_3H_6$	0.41	0.88	1.05[b]	40,602
Dehydrogenation	$n\text{-}C_6H_{14} \rightarrow 1\text{-}C_6H_{12} + H_2$	-2.21	-1.52	-	56,008
Polymerization	$3C_2H_4 \rightarrow 1\text{-}C_6H_{12}$	-	-	-1.2[c]	-
Paraffin alkylation	$1\text{-}C_4H_8 + iso\text{-}C_4H_{10} \rightarrow iso\text{-}C_8H_{18}$	-	-	-3.3[c]	-

[a]Data from Technical Data Book, Petroleum Refining, Division of Refining, American Petroleum Institute, New York, 1966 as cited by Wang [241].

[b]Data from Voge [11].

[c]Data from Ref. 275 as cited by Poutsma [276].

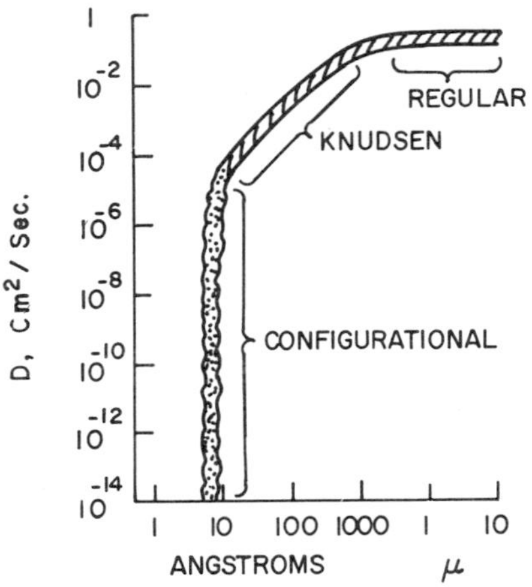

FIG. 36. Within the range of pore sizes—about 4 to 10 Å—encountered in a broad spectrum of zeolites useful in sorption and catalysis, we encounter a spread of diffusion coefficients (D) spanning 10 orders of magnitude. After Weisz [197].

may be considerably less endothermic. From another standpoint, Poutsma [276] observes that cracking reactions are driven by increases in entropy at high temperatures.

B. Location and Accessibility of Sites Active in Cracking

As observed earlier (Section VII-C), zeolitic pore-channel systems are uniform and of sizes characteristic of simple organic molecules. Thus diffusion in zeolites can be classified as lying at the extremity of the Knudsen region (Fig. 36), an area Weisz [197] has called the "configurational regime." Accordingly, diffusion rates may be sensitive to subtle structural features including polarity, polarizability, rigidity, and nominal critical dimension.

1. Locus of Catalysis in Faujasite

Weisz, Frilette, and associates presented definitive evidence that the major locus of catalytic activity in faujasite-type zeolites was within the intracrystalline pore volume [204]. This conclusion derived from a comparative study of the dehydration rates of sec-butanol over Linde 10X and 5A zeolites at high temperature and low conversion.

The specific rate constant (per unit volume) of the small pore zeolite (the 5 Å pores of which would not admit sec-butanol) was two to three orders of magnitude smaller than that of the faujasite 10X. These relative reactivities correlated with the accessible surface areas available for catalysis in the two zeolites: about 1 to 4 m^2/g for the external surface of the shape-selective 5A (assuming 1 to 5 μm crystal size) and 500 to 700 m^2/g for the internal surface of the larger pore 10X. However, at high reaction severities (particularly at high temperatures), reaction on the external surface of zeolites (the terminal faces of crystallites) must occur to some extent as evidenced by reports of formation of polymeric materials far larger in size than would be permitted egress through entry pores in both small pore (5A and chabazite) and faujasite zeolites [196, 278]. Possible roles for the matrix were discussed earlier (Section VII-D-7).

2. Shape-Selectivity in Small Pore Zeolites

Excellent reviews have been published by Weisz [197], Chen and Weisz [203], and more recently by Csicsery [279]. If we assume that the majority of catalytic sites are within the intracrystalline pore volume of zeolites, a highly selective type of catalysis, based on sieving or screening effects is possible. The earlier work of Weisz and co-workers [197, 203, 204] using zeolites with entry pore diameter near 5 Å showed selective cracking of linear paraffins but not branched paraffins (illustrated in Fig. 37), dehydration of primary but not branched alcohols, hydrogenation of only n-olefins in a mixture with isoolefins, and oxidation of n-butane but not isobutane. This work again demonstrated that catalysis in zeolites occurs within the internal cavities, with the restraint that only those molecules whose critical dimensions and molecular configuration are less than the dimensions of the entry window can participate.

A special type of reactant selectivity—best termed a cage or window effect—gives the impression of being related to a matching or fitting of the lengths of molecules and the zeolite cavity [279]. In cracking of n-$C_{22}H_{46}$ over the small pore zeolite erionite, two distinct maxima in size distribution of n-paraffin products were observed, with a well-defined minimum at C_8 [197]; diffusion coefficients from sorption experiments for a series of n-paraffins over a similar small pore zeolite (by R. L. Gorring) increased from $\sim 10^{-13}$ cm^2/sec at C_8 to $\sim 10^{-11}$ cm^2/sec at C_{12}—a striking variation [197]. In addition to parameters such as length and diameter, the various molecular oscillations may contribute to this behavior [197]. Still another type of shape selectivity—restricted transition state selectivity—may

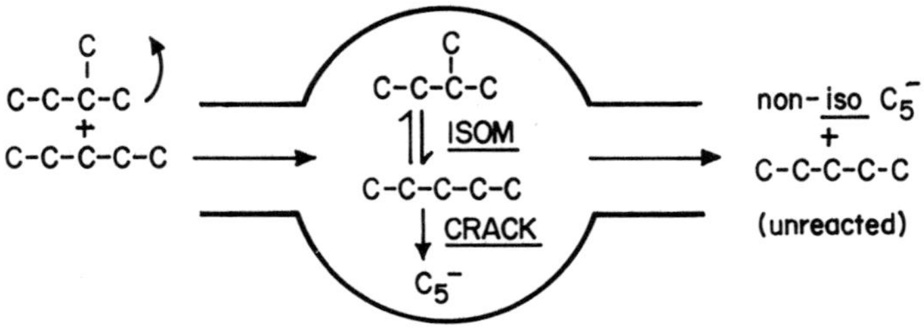

FIG. 37. Schematic visualization of reactant/product selectivity for cracking of n- and iso-C_5 paraffin mixture over a small pore (~ 5 Å) zeolite. The linear but not the branched isomer can pass through the entry pore, contact the catalyst active sites, and react. If bulky iso-structures are formed within the pores, they must be converted to less bulky species (by equilibration) or eventually degrade to form residue. After Venuto [280].

occur when the limited space of zeolite cavity prevents formation of the required transition state; a detailed perspective has been provided by Csicsery [279].

C. Basic Reaction Patterns

1. Radical/Thermal Cracking Mechanism

Excellent descriptions of the free-radical chain mechanism for homogeneous thermal cracking of paraffins (the Rice-Kossiakoff mechanism) have been provided by Greensfelder, Voge, and Good [180], Rabo and Poutsma [281], and Poutsma [276]. It is important to describe this pattern as a frame of reference, since a contribution from thermal cracking may sometimes accompany catalytic cracking at high temperatures (the thermal/catalytic interface).

The noncatalytic thermal reaction proceeds at 1022 to 1292° F via a radical chain of moderate length to produce a distinct and predictable product distribution, as shown in the data for cracking over quartz chips in Table 11. Initiation or activation of a paraffin (I) occurs via hydrogen abstraction to form the corresponding free radical (II) as in Eq. (1); R'· can be a hydrogen atom, methyl radical, etc. There are

$$\underset{\text{I}}{CH_3CH_2CH_2CH_2CH_2CH_2CH_2CH_2R} + R'\cdot \longrightarrow \underset{\text{II}}{CH_3CH_2CH_2\dot{C}HCH_2CH_2CH_2CH_2R} + R'H \qquad (1)$$

TABLE 11

Comparison of Products Observed in Thermal and Catalytic Cracking of n-Hexadecane at 932° F and 1 atm[a]

	Thermal, quartz chips	Activated carbon	Acidic, silica/ alumina/ zirconia[b]
LHSV[c]	0.05	10.0	10.0
Conversion, wt%	31.5	26.6	24.2
Selectivity, moles of product/ 100 moles cracked			
H_2	17	26	12
C_1	53	11	5
C_2	130	22	12
C_3	60	23	97
C_4	23	17	102
C_5	9	20	64
C_6	24	21	50
C_7	16	15	8
C_8	13	9	8
C_9	10	18	3
C_{10}	11	15	3
C_{11}	9	14	2
C_{12}	7	13	2
C_{13}	8	7	2 (C13–C14)
C_{14}	5	4	
C_{15}	-	14	-
Total hydrocarbon	378	223	358
Olefin content, wt%[d]			
C_3	78	-	-
C_5	99	-	-
108 to 165° F	90	-	-
Aromatic content, wt%			
108-210° F	1	-	-
210-257° F	3	-	-
iso-/n-butene ratio	0.07	-	-

[a]Data from Greensfelder, Voge, and Good [180].
[b]UOP-B cracking catalyst, 1/8 in. pellets, 346 m^2/g.
[c]Liquid hourly space velocity (fixed bed).
[d]Bromine number.

differences in activation energy for removing primary, secondary, and tertiary hydrogen atoms to form such radicals, with the corresponding C—H bond reactivities being about 1:3:10, respectively, at 1112° F [276, 281].

Radical II may immediately undergo β-scission to produce an olefin (III) and a radical of lower molecular weight (IV) (Eq. 2). As observed

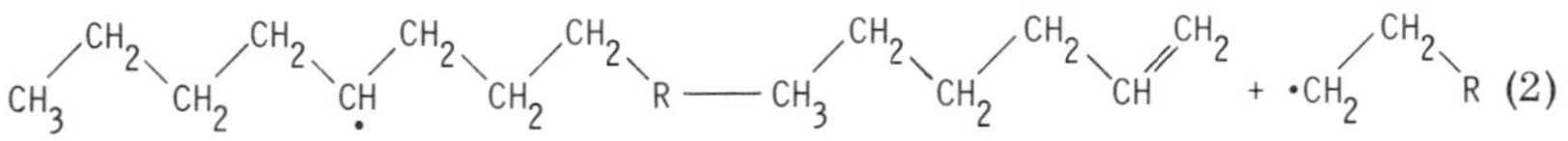

II III IV

by Rabo and Poutsma [281], if competition is available, the more stable olefin will be formed. At this point in the chain, the radical (IV) may either undergo further β-scission (rate $\propto k_\beta^{IV}$) or abstract hydrogen from substrate I (rate $\propto k_H^{IV}[I]$). The ratio $k_\beta^{IV}/k_H^{IV}[I]$ is $\gg$ 1 at temperatures and substrate pressures commonly observed in thermal cracking. If, however, a hydrogen atom (H•) or methyl radical must be extruded, hydrogen abstraction becomes competitive with β-scission [281].

After a succession of cracking/recracking events, the radicals are ultimately reduced to a product (Table 11) rich in C_1 and C_2 products and α-olefins, with little evidence for rearrangement of substrate products or intermediate radicals [180, 281]. The amount of butadiene is relatively high, but all other C_4's, particularly isobutane and isobutylene, are produced in very small amounts. Radical cracking characteristics are summarized in Table 12.

2. Radical-Type Cracking on Solid Surfaces

Voge and associates [180] observed rate enhancement when vapor-phase cracking of paraffins was conducted over silica gel, a supposedly "inert" solid; since "typical thermal" products were observed, radical pathways—with free-radical formation perhaps assisted by the solid surface—were proposed [180, 281]. Patterns over activated carbon (see Table 11) are also somewhat different from those over quartz chips [180]; activated carbon is a highly active, nonacidic catalyst, which gives a unique, quenched free radical-type reaction, perhaps acting as a sink to allow hydrogen atoms to combine with hydrocarbon-derived radicals, thereby preventing further fragmentation [180]. More paraffins than olefins are observed throughout the whole carbon number range, and there is less C_1 to C_3 than thermal. Although the surface structure of "activated" carbon is very different from that of

TABLE 12

Comparison of Major Features of Thermal and Catalytic Cracking[a]

Hydrocarbon type	Thermal cracking	Catalytic cracking
n-Paraffins, i.e., n-$C_{16}H_{34}$	C_2 is major product, with much C_1 and C_3, and C_4 to C_{16} α-olefins; little branching	C_3 to C_6 is major product; few α-n-olefins above C_4; much branching
Olefins	Slow double bond shifts and little skeletal isomerization; H-transfer is minor, and nonselective for tertiary olefins; only small amounts of aromatics formed from aliphatics at 932° F	Rapid double bond shifts, extensive skeletal isomerization; H-transfer is major and selective for tertiary olefins; large amounts of aromatics formed from aliphatics at 932° F
Naphthenes	Crack at slower rate than paraffins	If structural groups are equivalent, crack at about same rate as paraffins
Alkylaromatics	Cracked within side chain	Cracking next to ring is prominent

[a]Modified from Greensfelder, Voge, and Good [180].

the coke-on-catalyst produced during catalytic cracking, it is also apparent that they both are highly aromatic and have a high C/H ratio. Over pure alumina, which is weakly acidic, Greensfelder, Voge, and Good [180] observed a mixed cracking pattern, intermediate between that of strongly acidic oxides and activated carbon.

Thermal-Type Reactions over Zeolites

Reaction products reminiscent of radical pathways have also been observed over alkali metal-exchanged zeolites [281, 323]. Frilette, Weisz, and Golden [323], when cracking n-$C_{10}H_{22}$ over NaX, noticed significant rate enhancement and formation of a product rich in C_1, C_2, and C_3 fragments, low in C_4's, C_5's, and coke, and with evidence of double bond but not skeletal isomerization. The pattern appeard similar to a radical chain, but with heterogeneous modification, perhaps with surface participation in the chain initiation and rupture processes [323]. More recently, Rabo and Poutsma compared the reaction of n-C_6H_{14} and its isomers over KY and other alkali metal-exchanged zeolites with thermal cracking over quartz wool [281]. They observed a fivefold rate enhancement over thermal and some differences in product distribution. They were able to correlate their data with a radical mechanism by postulating zeolite-catalyzed double bond shifts subsequent to cracking, and that for a given radical (i.e., such as IV in Eq. 2 above), the ratio $k_\beta^{IV}/k_H^{IV}[I]$ is lower over KY than for thermal reaction. To help explain such a relative increase in the rate of hydrogen abstraction vs β-scission, Rabo and Poutsma [281] suggested a role for one property intrinsic to the three-dimensional structure of zeolites—a higher concentration of substrate (I) within its pore/channel network (compared to the vapor phase applied at the same pressure).

3. The Acid-Catalyzed/Carbonium Ion-Type Mechanism

Catalytic vs Thermal Cracking

In cracking over silica-alumina or other acidic materials, a significant rate enhancement and dramatically different product distribution is obtained. It is generally accepted that catalytic cracking proceeds through carbonium ion-type intermediates [154, 177, 178, 274, 281]. Product patterns from thermal and catalytic cracking of n-$C_{16}H_{34}$ at 932°F are compared in Table 11. The key difference is that catalytic cracking induces bond rupture selectively rather than randomly [177]. Greensfelder, Voge, and associates [154, 177, 178, 180, 282] cite rate accelerations for catalytic over thermal cracking on the order of 5 to 60, 10^3 to 10^4, and 5×10^2 to 4×10^3, for paraffins, olefins, and naphthenes, respectively. Gasoline products also differ markedly in molecular weight distribution and chemical composition [154]. Salient features distinguishing thermal and catalytic cracking are summarized in Table 12.

Characteristics, Stability, and Energetics of Carbonium Ions

Reactions of hydrocarbons over silica-alumina and acidic zeolites are similar to reactions catalyzed by a variety of strong acids in homogeneous media, where the intermediacy of carbonium ions has been clearly documented [196, 276]. Carbonium ions are electron-deficient, trigonal and planar, with a wide range of stabilities.

However, a spectrum of mechanisms is possible, depending on the extent of development of charge at the carbon atom; a carbonium ion may represent one limiting case, a polarized reactant another. In some cases, two catalyst functions might conceivably operate in a concerted mechanism [283]. The relative stability of carbonium ions (which reflects the selectivity of C—H bonds to form cations) is tertiary > secondary > primary > ethyl > methyl [241, 276, 284].

The difference in stability between methyl cation, for example, and the triphenylmethyl cation is striking [276]. Thus it is relatively unlikely that CH_3^+ and $C_2H_5^+$ cations will ordinarily appear as fragments in catalytic cracking, since reactions generating sec-propyl and tert-butyl ions are more favorable energetically by at least 26.5 kcal/mole [11]. A major mechanism of stabilization is hyperconjugation (with adjacent σ-bonds as in tert-butyl cation) and delocalization into associated π-electron systems as in benzyl and allyl cations [276]. An up-to-date summary of the properties of carbonium ion reactions is provided in the excellent review by Poutsma [276].

Illustrative Basic Reaction Patterns in the Catalytic Cracking Process

In addition to the basic "breakup" reaction of β-scission, other associated reactions (see Section IX-A) are observed, including paraffin and olefin isomerization, alkylaromatic isomerization and transalkylation, hydrogen transfer, as well as "build-up" type reactions such as polymerization, condensation, and coke formation [11].

For the case of a paraffin (I) the first or initiating event is the generation of a carbonium ion (II), say by hydride abstraction:

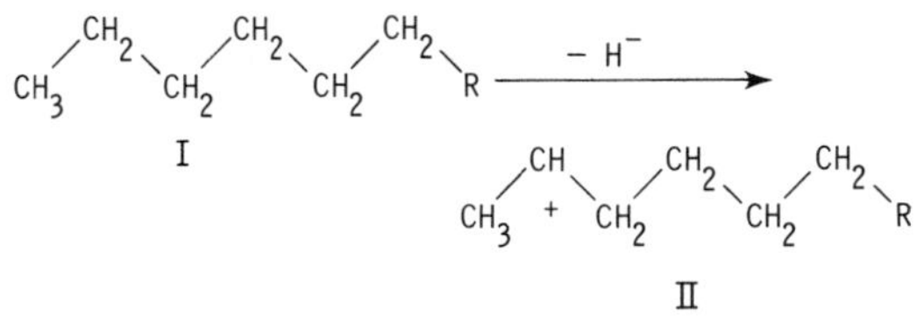

(3)

Carbonium ion II may undergo one of several basic reactions; (1) β-scission to form an olefin (III) and a smaller cation (IV) as in Eq. (4); (2) intramolecular positional (or skeletal) isomerization as in Eq. (5); (3) hydrogen transfer, where hydride is removed from a neutral molecule (V) to form a new carbonium ion (VII) and another neutral molecule (VI) [241] (in this case, the same as reactant (I) as in Eq.

(6); and (4) elimination to form an olefin (by donating a proton to a suitable base) as in Eq. (7) [274]:

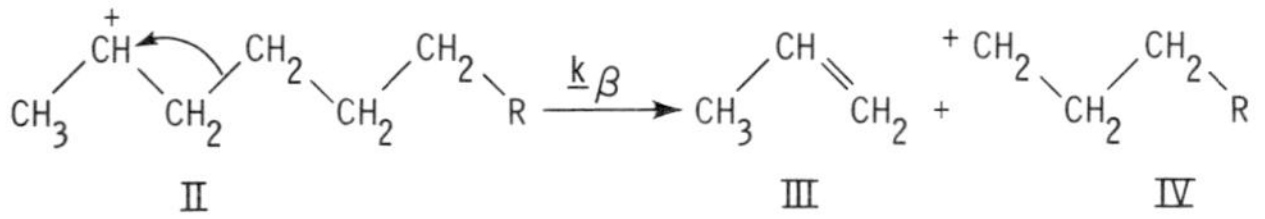

(4)

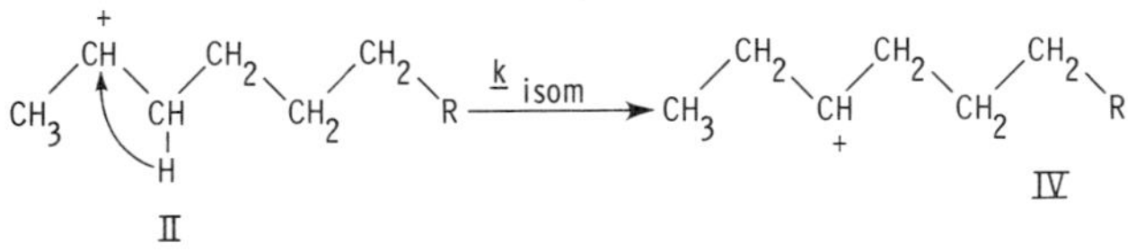

(5)

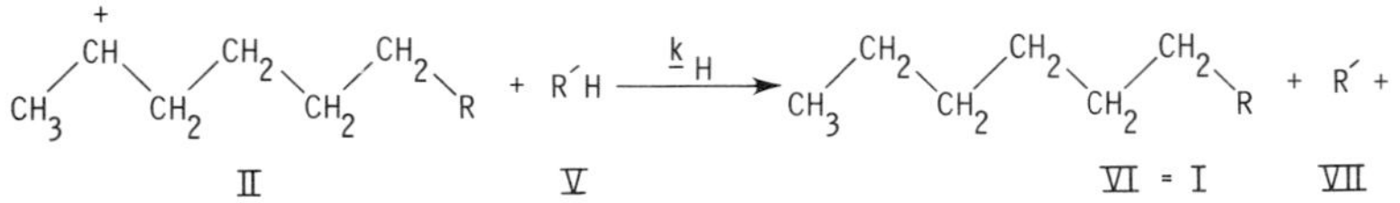

(6)

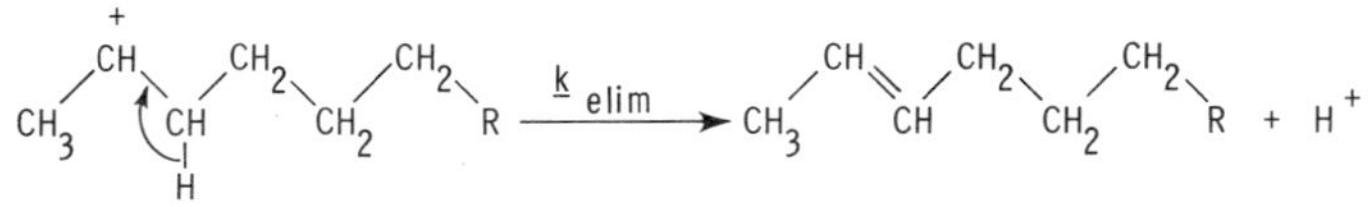

(7)

Equation (6) is a propagation via chain transfer, while Eq. (7) is a termination reaction [274]. Carbonium ions may also be quenched (destroyed) by attack of a suitable nucleophile [276]. Details are discussed below.

Initiation. Initiation via hydride ion abstraction from a paraffin has been frequently cited as a reasonable possibility [241, 248, 254, 274, 276, 281]:

$$A^+ + RH \longrightarrow AH + R^+ \quad (8)$$

This could be possible with direct attack via catalyst, i.e., if A^+ is a Lewis site on the surface [283] ($\geqq$Al + RH → R^+ + $\geqq$Al-H) or if A^+ is another carbonium ion (as in Eq. 6). A second mechanism postulates that a trace amount of olefin—from thermal cracking, etc.—is present in the paraffin, especially at high temperature; a proton donated from the catalyst generates a carbonium ion (reverse of Eq. 7), thereby setting up the reaction chain [11, 162, 180, 241, 248, 254,

274, 276, 283]. It has also been proposed that traces of oxygen or reactive carbon-on-catalyst could catalyze dehydrogenation of paraffin to form olefin [11].

A role for olefins as initiating species was clearly demonstrated by Weisz [248] for cracking of the relatively unreactive n-butane over H-mordenite zeolite at low temperatures (400 to 480° F). When small amounts of butene were added to butane or when butane was pre-exposed to a reactor zone containing a Pt/nonacidic Al_2O_3 dehydrogenation catalyst at temperatures high enough to establish the equilibrium n-$C_4H_{10} \rightleftarrows C_4H_8 + H_2$, significant conversion of n-C_4H_{10} was observed. Without either the added olefin or the presence of a dehydrogenating component, no cracking occurred at these conditions.

Another possibility for initiation is removal of hydride ion via direct attack of proton from a Brönsted site on the catalyst surface [11, 180, 276, 283]:

$$RH + H^+ \longrightarrow R^+ + H_2 \tag{9}$$

This would lead to formation of traces of H_2 gas. There are some convincing analogies in the literature for various Brönsted "super acids" attacking paraffins in this manner [276]. Poutsma [276] also suggests a possible initiating route where a proton from a Brönsted site attacks a paraffin chain, breaking a C—C bond (Eq. 10), perhaps via a pentacoordinate homolog of CH_5^+.

$$CH_3-CH_2-CH_2 \,\vdots\, CH_2R + H^+ \longrightarrow CH_3-CH_2-CH_3 + {}^+CH_2R \tag{10}$$

<u>β-Scission</u>. β-Scission is heterolytic splitting of a C—C bond [241] with generation of shorter chain olefin molecules [274]. Its reverse, polymerization, is not favored by thermodynamics at temperatures typical of catalytic cracking [11, 276]. If reactant structure permits alternatives in formation of carbonium ions, the more stable cations participate in the reaction events to a far greater extent than the less stable ions [241]. Thus the first and second C—C (σ-) bonds of an aliphatic chain are relatively inert to the action of the catalyst, i.e., there is higher activation energy to form methane or ethane, and very small amounts of these are produced at ordinary severities [177, 253, 281]. However, the third C—C bond from the end is relatively easily broken (i.e., to form C_3H_6 as in Eq. 4), and there may even be a very slight preference for β-scission of bonds further toward the center of the molecule [11, 177, 253, 276, 281, 284].

If a newly formed carbonium ion is primary (Eq. 4), i.e., β-scission has resulted in a less stable system, there will be a tendency to form a more stable configuration by 1,2,-hydride shift as in Eq. 5 [274]. After a cascade of multiple β-scissions, the derived residual carbonium ion may have become so small that further cleavage would result in formation of a very unstable carbonium ion, i.e., CH_3^+; in such a situation, hydride transfer—with initiation of a new chain—may be favored [237, 276].

It is important to emphasize that all the ionic reactions depicted in Eqs. (4) to (10) are occurring with adsorbed carbonium ions at anionic sites on the surface of the catalyst. Thus, in Eq. (6), when hydrogen transfer occurs, the adsorbed species (II) is promptly desorbed (VI), and is replaced by a new adsorbed species (VII) [253, 284]. The probability of recracking a residual adsorbed hydrocarbon fragment before hydrogen transfer and desorption greatly increases at higher temperature [253]. Stated differently, β-scission is more rapid than hydrogen transfer (and isomerization) at elevated temperatures. Also, as observed by Poutsma, β-scission should become increasingly important at higher temperatures for thermodynamic reasons [276].

Isomerization. Because carbonium ions can undergo facile 1,2-migration of hydride, alkyl, or aryl groups, a considerable amount of isomerization—primarily skeletal and usually non-equilibrium—takes place during cracking [11, 241, 276, 281, 284]. This leads to a product configuration with high ratios of branched to normal products [276], and is responsible for the high yields of isobutane as the primary product from cracking of long straight-chain paraffins [11]. These cationic isomerizations are far more rapid than the radical type and lead to conversion of primary carbonium ions to more stable species prior to β-scission [281]. The relative rates of isomerization and β-scission (k_{isom}/k_β) govern the extent to which a carbonium ion system can achieve a quasi-equilibrium (in terms of possible skeletal isomers) prior to drain off by cracking [276].

There is a tendency to approach equilibrium with aromatics and olefins under cracking conditions (Table 10); the olefin isomerizations are usually secondary reactions of olefins formed by primary cracking events [11, 284]. Isomerization of paraffins under typical catalytic cracking conditions is not an important reaction [11]; n-C_4H_{10} is relatively inert and not easily isomerizable to iso-C_4H_{10} [284]. Aromatic transalkylation can also occur [11].

Hydrogen Transfer. Hydride transfer appears to be a dominant process underlying the profound hydrogen redistribution observed with zeolites. This will be discussed in detail below.

"Buildup Reactions." Polymerization, condensation, and cyclization reactions—generally associated with the genesis of coke—accompany catalytic cracking [11, 241]. At low temperatures and higher pressures (where equilibria as well as surface concentrations of adsorbed olefin are more favorable), cracking catalysts are active for cationic olefin polymerization [11, 196, 276]. However, because of thermodynamic factors, polymerization/condensation reactions become disfavored with respect to their reverse—β-scission—at higher temperatures [11, 230, 276]. However, even under conditions where thermodynamics limits equilibrium concentrations of simple polymers, polymerization often constitutes the first stage in the formation of coke from olefins [11]. Voge [11] also suggests that cross-linking of some aromatics, with elimination of H_2 gas, can sometimes occur under cracking conditions.

Dehydrogenation. Direct dehydrogenation is not characteristic of cracking unless metals poisons are present (see Section VIII-C-3 and below). With hydroaromatics like tetralin (naphthene ring adjacent to aryl ring), dehydrogenation (as distinct from hydride transfer) to form H_2 may occur [11].

Structure/Reactivity Relationships for Paraffins, Olefins, Naphthenes, and Alkylaromatics in Carbonium Ion-Type Cracking

General patterns for cracking of various classes of pure hydrocarbons are shown in Fig. 38. Such information provides a fundamental base for understanding the cracking of gas oil. In general, for a given carbon number, the following order of reactivity was observed [11]: olefins (not shown in Fig. 38) > aromatics with C_3 or higher alkyl substituent > naphthenes > polymethylaromatics > paraffins > aryl rings. Aromatic nuclei are highly resistant to cracking. All classes showed rate increases with increasing molecular weight [11].

Paraffin Reactivity. Voge [11] suggested that the increase in cracking rate with carbon number in n-paraffins was too great to ascribe to the greater number of C—C bonds (or secondary C—H bonds); he proposed that more extensive adsorption might favor a greater rate of activation/carbon atom [11]. For cracking of a series of C_8 to C_{10} n-paraffins over REHX zeolite (Table 13), Nace [254] also observed an increase in cracking rate constant (k) with carbon number—up to a maximum of C_{16} followed by a decline. Note in Table 13 that coke-on-catalyst also increased continuously with carbon number. It appears that deactivation of the catalyst eventually more than offset the gain in reactivity afforded by longer chain length; if, however, coke levels were the same at the instant of the conversion

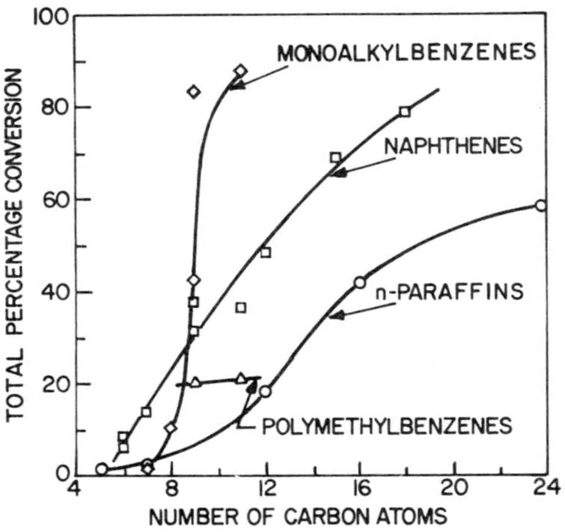

FIG. 38. Major reactivity trends for catalytic cracking of representative classes of hydrocarbons over silica-alumina-zirconia catalyst at 932° F. Curves do not represent true intrinsic reactivities because average catalyst activities over the 60-min run period were not constant due to different rates of coke formation for the various reactants. After Voge [11].

determination, reactivity would probably have continued its monotonic trend with increasing carbon number [254]. In a line of reasoning similar to that of Voge [11] above, Katzer [283] suggested that the rate enhancement observed by Nace [254] with increasing carbon number did not relate to a greater number of secondary carbons alone, but to a higher surface concentration of the longer-chain paraffins within the zeolite pores [285].

Greensfelder et al. [180], in early definitive studies over amorphous catalysts, estimated that tertiary C—H bonds were 10 times as reactive as secondary and 20 times as reactive as primary. That is, the reactivity ratio of primary: secondary: tertiary C—H bonds is 1:2: 20 [11, 274]. Nace's data [254] for cracking of tetramethylpentadecane over an acid zeolite catalyst (Table 13) demonstrate the increased reactivity of highly branched paraffins. This iso-C_{19} alkane is nearly five times as reactive as n-$C_{18}H_{38}$—even with a higher coke buildup on the catalyst, probably reflecting the plurality of tertiary configurations on the chain. The iso-C_{16}, heptamethylnonane, with

TABLE 13

Cracking of Various Paraffins over REHX at 900° F[a, b]

	Rate constant, k	Wt% C on catalyst	Moles product/ mole cracked
n-C_8H_{18}	36	0.25	2.0
n-$C_{12}H_{26}$	660	0.70	2.5
n-$C_{14}H_{30}$	984	1.0	2.8
n-$C_{16}H_{34}$	1000	1.4	3.0
n-$C_{17}H_{36}$	738	1.7	3.1
n-$C_{18}H_{38}$	680	2.2	3.2
Tetramethylpentadecane (an iso-C_{19})	3300	2.8	3.1
Heptamethylnonane (an iso-C_{16})	164	0.5	3.5

[a]Instantaneous values at 900° F.
[b]Data from Nace [254].

most of its branching in quaternary groupings, shows only a fraction of the activity of n-$C_{16}H_{34}$ at one-third of the coke loading. The retardation by quaternary groups [274] may, in part, reflect the absence of tertiary groups.

Olefin Reactivity. Olefins crack at a much faster rate than paraffins over amorphous [11, 274] and zeolite [254] catalysts and are subject to rapid isomerization. This is in accord with the protonation pathway for initiation of carbonium ion activity discussed earlier.

Naphthene Reactivity. Naphthenes, with their greater number of secondary carbon atoms—or tertiary as in decalin or substituted cyclohexanes—cleave more readily than the corresponding straight chain aliphatics [180]. Similarly, cycloolefins undergo β-scission considerably more rapidly than the corresponding cycloparaffins [274].

Aromatic Reactivity. The main feature in cracking of alkylaromatics is the great stability of the phenyl ring and the relative ease of chain removal by dealkylation [274]—via a carbonium ion-type mechanism [196, 276, 280]. The ease of side-chain cracking by β-scission increases when the alkyl group is long [11, 237, 274].

4. The Catalytic Effect of Metals

The poisoning effect of metals contaminants upon activity and selectivity in cracking was discussed in detail earlier (Section VIII-C-3). It is well known that nickel catalyzes dehydrogenation and condensation reactions—including formation of carbon and hydrogen—at high temperatures [286]. Indeed, even under the conditions of hydrogenolysis, Taylor, Sinfelt, and Yates [287] suggested that ethane formed an unsaturated surface radical on nickel supported on kieselguhr via an initial endothermic dehydrogenation—attesting to nickel's dehydrogenative power. Aromatics have been generated from C_6 and C_7 naphthenes [286], and small amounts of olefins and aromatics have been reported from reaction of paraffins over NiX zeolite [288]. Adsorbed carbon residue was also reported by McKee [289] in a study of ethylene decomposition on clean, unsupported nickel powder at low temperature (392° F). Nickel-catalyzed reactions such as these probably account for most of the "contaminant" coke and hydrogen formation observed in metals-poisoned cracking systems.

Despite the known propensity for selective attack of nickel catalysts on the terminal C—C bonds in alkanes—the "methane snipping" (α-cracking) reaction so prominent in hydrogenolytic studies [290]—only minor increases in quantities of methane seem to be observed in metals-poisoned catalytic cracking situations [260].

5. Selectivity Differences between Zeolite and Amorphous Catalysts: The Dominance of Hydrogen Transfer

Oblad [13] has observed that the activity of zeolites is generally similar in type to that of amorphous silica-alumina, but that there are differences in both quantity and quality of acidity (i.e., acid strength distribution) that may account for the activity/selectivity differences. Further, as noted by Weisz [101], the observation that the energy of activation for cracking ($\sim$30 kcal/mole) is nearly identical for a number of zeolites and silica-alumina infers that their mechanistic pathways are, at least, not drastically different. Part of

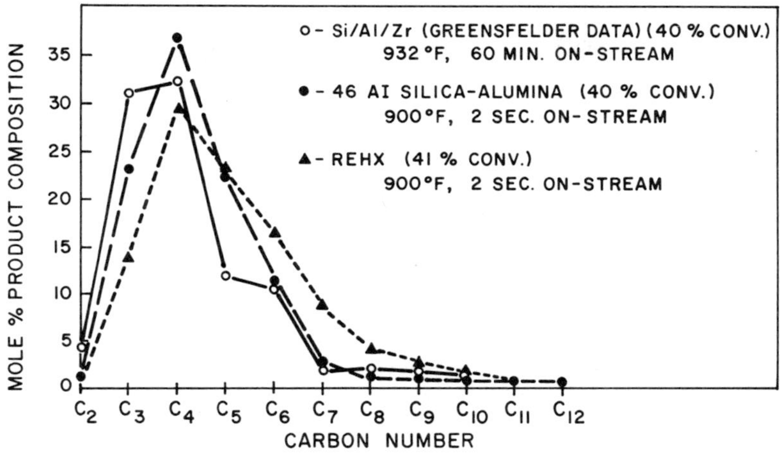

FIG. 39. Comparison of product distributions from catalytic cracking of n-hexadecane over REHX zeolite and amorphous silica-alumina (fixed bed microreactor). Data are from Nace [237]. Note the rather good agreement with the earlier cracking data from Greensfelder et al. [180].

the activity difference (discussed in Section VII-D-1) may relate to a site density in zeolites about 10 to 100 times greater than that in silica-alumina [283]. Part may also relate to a significantly greater effective concentration of reactant molecules—strategically located near active sites—reflecting the influence of the zeolitic channel system with pore sizes close to molecular dimensions [283, 285]; Katzer [283] estimates a 50-fold concentration advantage in zeolites over silica-alumina. Finally, part may reflect the activating effects of the high [212] zeolite electrostatic field potentials (Section VII-C) on hydrocarbons, and C—H bonds in particular.

Product Distribution Shifts: Zeolite vs Amorphous

Nace [237], in an excellent analysis of cracking patterns obtained over crystalline and amorphous catalyst, noted that with REHX zeolite there was better selectivity for medium molecular weight liquid products, i.e., gasoline, than with silica-alumina. As shown in Fig. 39, there is substantially less gas (C_2 to C_4) and more gasoline (C_5 to C_{11}) with the zeolite, giving a significantly lower ratio of moles of

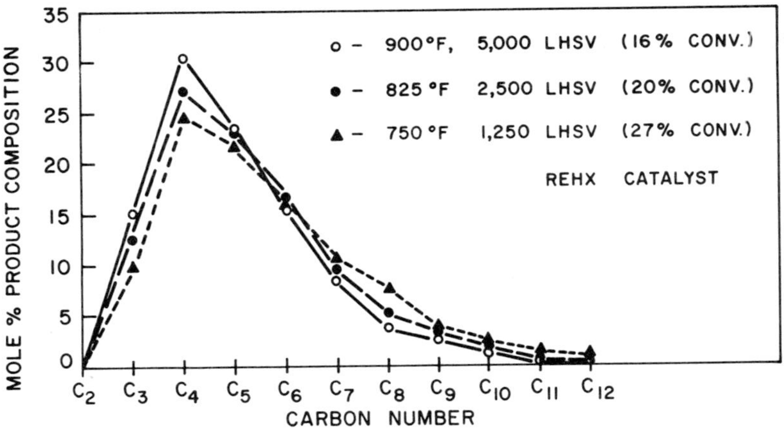

FIG. 40. The effect of lowering temperature on product distribution from cracking n-hexadecane over REHX is similar to that observed in going from amorphous to zeolitic catalyst, i.e., the ratio $k_\beta/k_H[RH]$ decreases at lower temperatures. The chain process of β-scission is stopped at an earlier stage at low temperature. Data are from Nace [237].

product/mole of feed cracked. Nace [237] ascribed this to a decrease in the rate of C—C β-scission relative to the rate of hydrogen transfer for the intermediate carbonium ion. That is, the ratio of $k_\beta/k_H[RH]$ (where RH is a hydrogen donor) is smaller with the crystalline aluminosilicate [237], allowing the hydrogen transfer reaction to proceed more nearly to completion than with conventional silica-alumina [190]. Thus cracking of hydrocarbon chains is intercepted (terminated) at higher molecular weight by hydrogen transfer of hydride to the carbonium ion (as in Eq. 6). Also, as observed by Thomas and Barmby [190], the ratio of the rate of hydrogen transfer forming volatile aromatic products to the rate of reactions which generate coke is greater with molecular sieves than with amorphous catalysts.

It was known with amorphous catalysts that at temperatures around 900 to 930° F, hydrogen transfer occurred somewhat more slowly than cracking with gas oils, but that its rate was relatively faster at lower temperatures [11]. A similar—but more pronounced—pattern exists with zeolites, as can be seen in Fig. 40.

Structural Considerations Relating to Zeolitic Selectivity for Hydrogen Transfer

The tendency of microporous zeolitic solids to sorb hydrocarbons and thereby concentrate reactants would tend to favor bimolecular (probably second order) hydrogen transfer reactions (rate $\alpha k_H[R^+][RH]$ [276, 283].

Selectivity in zeolites may be sensitive to the distribution of acid site strengths [13, 283]; indeed, silica-alumina appears to have a higher fraction of total acidity in the strong acid range [13]. There is also a distribution of site energies in zeolites. Mikovsky and Marshall [291], postulating random aluminum siting in the faujasite lattice, found four degrees of acid strength for associated protons. "Strong" acid sites, type -0, comprised only about 20% of the total, the remaining types -1, -2, and -3 being classified as "weak."

Moscou and Moné [292], in correlating cracking selectivity of zeolites with strength of acid sites (titration with indicators), observed that the strongest acid sites gave undesirable selectivity toward coke and gas. Katzer [283] proposes that the presence of weaker sites in zeolites than in silica-alumina may be related to their selectivity advantage, particularly in terms of less active cracking relative to hydrogen transfer; the very strength of the acidity of silica-alumina might tend to favor tight binding of an organic cation to the surface, mitigating against hydrogen transfer followed by desorption, and favoring a continued succession of cracking events and coke formation from localized species.

Olefins, Hydrogen Transfer, and the "Gasoline Stabilization" Mechanism

Thomas and Barmby [293] point out that the hydrogen transfer process in aluminosilicates resembles a high temperature variety of "conjunct polymerization" in which olefins, in the presence of a strong acid such as H_2SO_4 or H_3PO_4, are transformed into hydrogen-rich paraffins and hydrogen-deficient products. Presumably, at elevated cracking temperatures, a similar process occurs at least initially which, via a complex sequence of reactions (including cyclization), leads to paraffins and aromatics which are desorbed into the product stream.

Thus the transfer of hydrogen from some source to olefins is highly characteristic of catalytic cracking [11]. Olefins are, in fact, the most reactive components in gasoline [294] and are the largest contributors to secondary cracking reactions [248]. Some of the typical hydrogen-transfer reactions with olefins as reactants include olefin-naphthene, olefin-olefin, or cycloolefin/cycloolefin reactions [294]:

$$3C_nH_{2n} + C_mH_{2m} \longrightarrow 3C_nH_{2n+2} + C_mH_{2m-6} \qquad (11)$$

olefins + naphthene ⟶ paraffins + aromatic

$$4C_nH_{2n} \longrightarrow 3C_nH_{2n+2} + C_nH_{2n-6} \qquad (12)$$

olefins ⟶ paraffins + aromatic

$$3C_mH_{2m-2} \longrightarrow 2C_mH_{2m} + C_mH_{2m-6} \qquad (13)$$

cyclo-olefins ⟶ naphthenes + aromatic

Thus some of the hydrogen transferred to olefins comes from compounds that are, in essence, dehydrogenated to aromatics, but with little if any transfer of molecular hydrogen, H_2 [11]; some of the hydrogen comes from molecules that are condensed to form coke [11]; i.e., with both amorphous and zeolitic catalysts, part of the aromatics formed remains adsorbed on the catalyst and deactivates it [190].

Hydrogen transfer accounts for the very significant quantities of isobutane and octane-enhancing isoparaffins formed during catalytic cracking [11]. Such high iso/n ratios in paraffin products derive from the hydrogen transfer to tertiary olefins that occurs orders of magnitude faster than with n-olefins; this process, together with skeletal isomerization of olefins, generates an abundance of isoparaffins that may greatly exceed equilibrium expectations.

Of course, a major effect of hydrogen transfer is to stabilize or preserve gasoline range olefins formed by primary cracking of gas oil, prior to their drain-off by secondary cracking [13, 197]. The dynamics of this reaction are illustrated in Fig. 41 for a reaction such as depicted in Eq. (11); Weisz [197] describes experiments (by S. K. Aditya) for this reaction which conform almost exactly to this stoichiometry.

Experimental data showing trends in P, O, N, A composition for C_6^+ gasoline vs extent of conversion [99]—and demonstrating the "stabilization" phenomenon discussed above—is shown in Fig. 42. Note that over the temperature range of 900 to 1100° F, olefins and naphthenes decreased; correspondingly, paraffins and aromatics generally increased, especially at the higher temperatures. This redistribution of hydrogen is particularly facile with hydrotreated

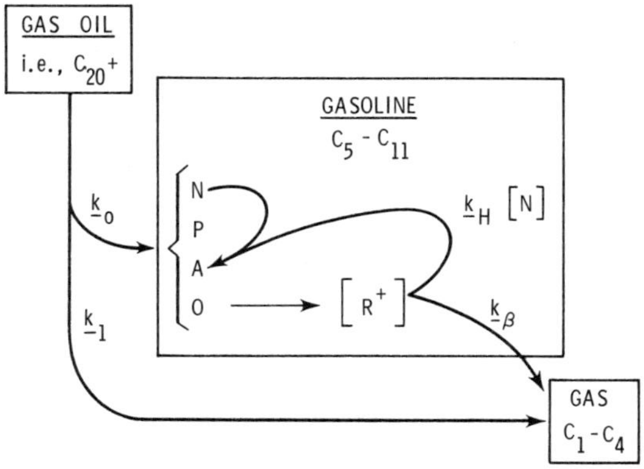

FIG. 41. Integration of an early kinetic scheme by Weekman and Nace [41] with concepts advanced by Nace [237] and Weisz [197]. Preservation of molecules in the gasoline range is illustrated for the facile hydrogen transfer reaction O + N → P + A. A gasoline range carbonium ion, R^+, derived from olefin, O, is converted by reaction with naphthene, N, to more refractory paraffins and aromatics. This happens before R^+ can degrade by further cracking to short-chain gaseous fragments. k_0 and k_1 represent primary gas oil cracking paths.

stocks [101, 151]. The data in Fig. 42 infer that the rate of hydrogen transfer (but not necessarily $k_H[RH]/k_\beta$) is increasing with temperature. The large (18 vol%) increase in aromatics (Fig. 42D) with the associated decrease in paraffins and naphthenes at 1100° F reflects, in part, a "concentration effect," as saturated species are selectively cracked away at the higher severity. Further, within the class of aromatics, the ratio of naphthalenes/tetralins (at the upper end of the gasoline boiling range) greatly increased with temperature. At 1000° F, although there were large changes in the relative proportions of gasoline P, O, N, A components over the conversion range 74 to 84 vol%, measured octanes varied only slightly. This suggests that compensation effects were operative, i.e., one class of high octane components (olefins) was replaced by another (aromatics).

Locus of Catalysis in Zeolite Catalysts vs Hydrogen Transfer

Thomas and Barmby [190], in assessing selectivity differences between zeolite and amorphous catalysts, proposed that the rate of

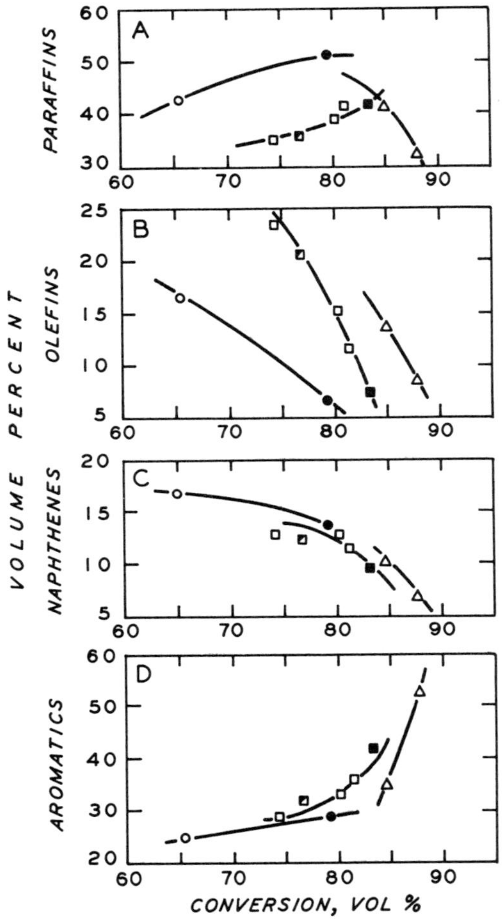

FIG. 42. Trends in molecular composition (P, O, N, A analysis) vs conversion for C_6^+ gasolines obtained in short contact time riser cracking of hydrotreated heavy vacuum gas oil over a fluid zeolite catalyst; O, □, Δ = 900, 1000, 1100° F. Data from Owen, Snyder, and Venuto [99].

diffusion of gas oil molecules through zeolite entry pores was too slow to account for the rate of cracking. They assumed that in commercial composite cracking catalysts, which contained small zeolite crystallites embedded in an amorphous matrix (as in Fig. 13), gas oil molecules were too large to enter the zeolite cavities, instead cracking on

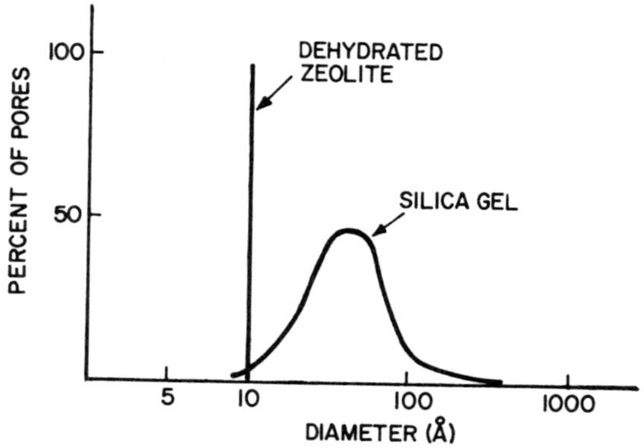

FIG. 43. Distribution of pore sizes in microporous adsorbents; modified after Breck [183]. Pore size distribution of silica-alumina may be on the order of 30 to 80 Å [190].

the external surface area of the zeolite crystallites or in the amorphous matrix. The data in Fig. 43 contrast the uniform pore dimension of faujasite with the broad pore size distribution of an amorphous material typical of that employed as a matrix.

They then postulated that the lower molecular weight cracked fragments (i.e., gasoline boiling range) could then enter the zeolite cavities without having to compete with large gas oil molecules for active sites. Stabilizing hydrogen transfer reactions could then occur within the vastly greater intrazeolitic surface area.

Weisz and associates [101, 248], while agreeing that hydrogen transfer was critical, found no need to invoke the proposed [190] separate involvement of external and internal surfaces. The improvement in gasoline selectivity and other differences in product distribution could be rationalized merely by a higher ratio for rate of hydrogen transfer vs β-scission ($k_H[RH]/k_\beta$) per unit volume of zeolite alone.

6. Kinetics and Chemistry of Coke Formation

This discussion will focus on dynamics of "catalytic" coke formation, as contrasted to "contaminant" coke, "additive carbon," and "cat-to-oil" coke, as defined in Section VIII-C-3.

Nature of Coke

Haldeman and Botty [295] conducted comprehensive studies on the composition of coke (carbon) deposits on silica-alumina cracking

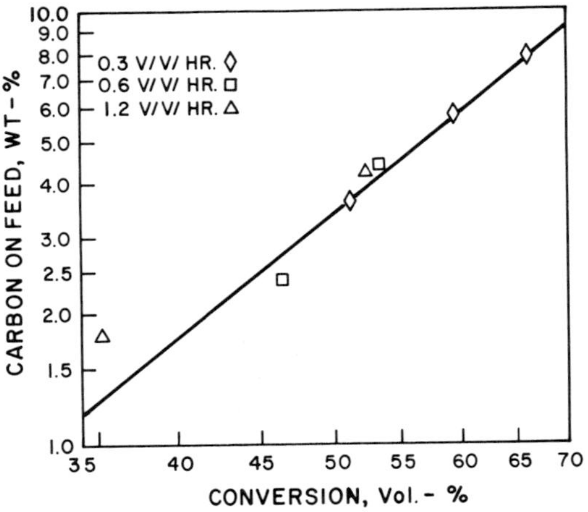

FIG. 44. Carbon yield vs feedstock conversion (log-log plot) for cracking of East Texas Gas Oil over synthetic amorphous catalyst in a fixed bed. After Voorhies [166].

catalysts. Using x-ray diffraction as well as other techniques, they determined that two major types of carbon structures were present as highly dispersed phases within the catalyst pores. The largest part consisted of pseudo-graphitic or turbostratic, random-layer lattices (individual layers similar to those of graphite but with no three-dimensional or true crystallinity); the C/H ratio of this fraction was 2.0 to 2.6. The remaining portion comprised poorly organized (sometimes substituted) polynuclear aromatic systems. Various tracer studies (^{14}C) on amorphous [104, 296] and zeolitic [297] catalysts have shown negligible exchange of labeled hydrocarbons with coke.

Kinetics of Coke Formation

For a given catalyst, feedstock, and temperature, good correlation between feedstock conversion and carbon yield (on feed) is generally observed [166]. Plots basically similar to that in Fig. 44 have been obtained with data from fluid catalytic cracking and with zeolite catalysts.

Voorhies [166] recognized at an early stage that although the degree of carbon formation in catalytic cracking was dependent on the type of catalyst, feedstock composition, and operating conditions, carbon nevertheless deposited on the catalyst in a fundamentally uniform manner with time. This led to his finding, from analysis of a

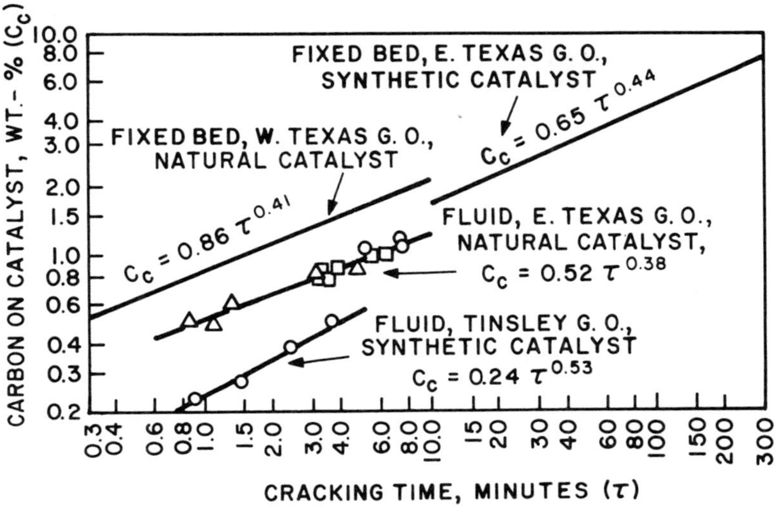

FIG. 45. Adaptation of original data of Voorhies [166] for the formation of carbon on catalyst vs catalyst residence time for fixed and fluidized bed cracking catalysts. τ in the equation $C_C = A\tau^n$ averages about 0.44 for these amorphous catalysts; A ranges from about 0.2 to 0.9 for the various feedstocks. All catalysts were very well stripped, so that carbon formation represents substantially catalytic carbon alone.

large quantity of data, that the wt% carbon laid down on the catalyst (C_C) was logarithmically proportional to its time of exposure to feed since regeneration. This quantitative relationship is the now classical Voorhies equation [166], the "coke clock":

$$C_c = A\tau_c^n \tag{14}$$

A is a coefficient that varies with feedstock properties, catalyst, and temperature [166], τ_C is the catalyst residence time, while n is also a constant. Log-log plots demonstrating this relationship for several different catalysts and gas oils are shown in Fig. 45. A is a measure of the coking tendency of a particular feed [109, 110]; the value of A increases with the concentration of high molecular weight aromatics [109].

In a detailed study of cracking behavior of 16 feedstocks of various types over zeolite catalysts, Nace, Volz, and Weekman [109]

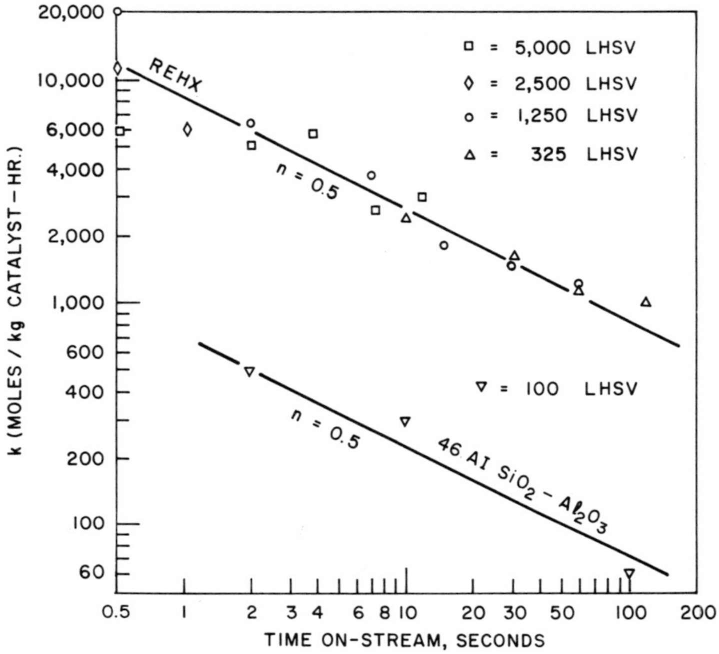

FIG. 46. Plot of k from the exponential decay function $k = k_0\tau_c^{-n}$, vs time on stream, where τ_c is in seconds, n is a (decay) constant, and k is a pseudo first-order rate constant that is used as a measure of the activity of the catalyst at a given instant. Data from Nace [237].

found that the Voorhies constant, n, varied over a range of 0.12 to 0.30, averaging about 0.21. This is characteristically much lower than the values of n obtained with amorphous catalysts, as in Fig. 45; values for A ranged from about 0.2 to 0.8 [109]. Aging data for both zeolitic and amorphous catalysts are displayed in a different format in Fig. 46. Voorhies [166] and Levinter [298] both cite the existance of a limiting (maximal) degree of coking beyond which no further coke formation can occur, probably due to blockage of the remaining pores and accessible surface. It will be recalled that the kinetics of coke burning were discussed in Section VIII-F-1.

The Chemical Pathways Leading to Coke

The great preponderance of data focuses on reactions of aromatics [11, 38, 41, 110, 123, 283, 299] and olefins [175, 253, 283-285,

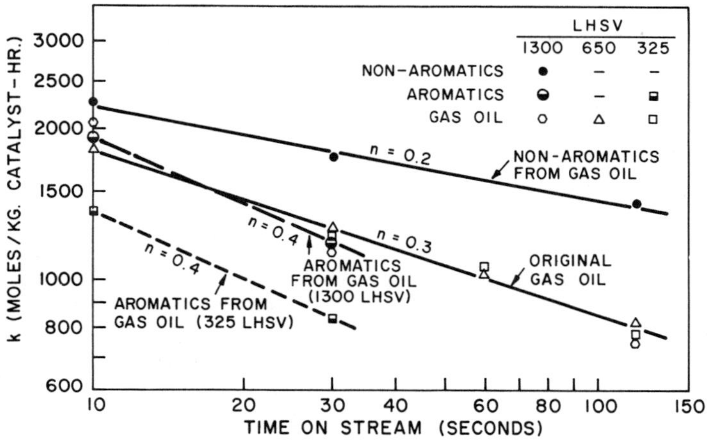

FIG. 47. Plot of k (from exponential decay equation, $k = k_o \tau_c^{-n}$) vs catalyst residence time for Light East Texas Gas Oil (LETGO), and an aromatic extract and a nonaromatic fraction from the same gas oil; fixed-bed, REHX zeolite catalyst, 900° F. Data from Nace [123].

298, 300, 301] as leading to coke formation. It might be more appropriate to state that coke formation over acidic solids derives largely from transformations of aromatic hydrocarbons—either initially present as reactants or generated from other precursors (i.e., olefins) by secondary processes during the catalytic cracking process [38]. There is some indication that increased catalyst site density and acid strength may favor increased rates of coke formation [283]; it is also possible that a dual-site mechanism (with closely spaced sites) may be necessary to rationalize coke formation via certain hydrogen-transfer reactions.

Factors Related to Aromatics as Primary Reactants. Deposition of coke on catalyst—with a consequent decline of catalytic activity—is probably a primary reaction if the feed contains high molecular weight, multi-ring (i.e., polynuclear) aromatic hydrocarbons [38, 41, 123, 299]. Volz, Nace, and Weekman [110], for instance, quantitatively relate the catalyst decay constant k (defined in Fig. 47) to the aromatic/naphthene ratio in virgin gas oil. Similarly, Nace [123] observed that the rate of catalyst decay for an aromatic fraction extracted from Light East Texas Gas Oil (LETGO) was considerably faster than the nonaromatic portion or the original LETGO (Fig. 47). The decay constant k for the aromatic concentrate was also sensitive to space velocity. The amount of coke deposited on the catalyst during a 10-sec exposure was about three times as high with the aromatic fraction as with the nonaromatic portion; in addition, with the aromatic extract, higher molecular weight (C_{18} to C_{19} components) increased from 1.5 to 7.9% during the run, suggesting buildup of coke precursors by condensation reactions [123].

Appleby, Gibson, and Good [38] also showed that coke formation was much higher in cracking a recycle oil containing a high level of polycyclic aromatics than with the same stock after catalytic hydrogenation. They noted a distinct correlation between coke-forming tendencies of aromatics and their basicity constants (Fig. 48)—reflecting, most likely, their propensity for chemisorption on (electron-withdrawing) eatalyst surfaces. Appleby, Gibson, and Good [38] postulated that such adsorbed aromatics might form reactive ionic species that could couple or condense with other aromatics and unsaturated molecules. This would lead to a "growing" solid or semi-liquid "polymer" and, via dehydrogenation/condensation reactions, ultimately form aggregates of polynuclear aromatics exhibiting the turbostratic graphite structure of coke. They also suggested that there might be a synergistic effect on coke formation if olefins as well as aromatics were present, and that the coke-formation process would probably be accelerated by the carbonium ion reactions that occur in catalytic cracking.

Factors Related to Olefins as Reactants Ultimately Leading to Coke. Hydrogen transfer reactions, in an extension of the reasoning developed earlier (Section IX-C-5), are highly important contributors to coke formation [170, 175, 178, 283, 284, 300]. Here, olefins act as acceptors or sinks for hydrogen and are saturated, while the olefinic or perhaps naphthenic donor molecules remain behind as adsorbed carbonium ions or unsaturates. After a succession of such hydride-transfer events, increasingly hydrogen-deficient unsaturated or aromatic species remain immobile and strongly adsorbed on the catalyst surface, ultimately forming coke. This type of sequence is visualized in Fig. 49 using the hydrogen redistribution reaction of the simple olefin ethylene over an acidic zeolite [280] as an example. It appears that cationic polymerization, cyclization, cracking, and dehydrogenation-type processes—accompanied by a cascade of intermolecular hydride transfers and light paraffin extrusion—are operative in the system described in Fig. 50 [196, 202, 276, 280, 285, 301].

This type of hydrogen-transfer process begins at temperatures as low as 150 to 200° F with ethylene in zeolites [301]. Shephard,

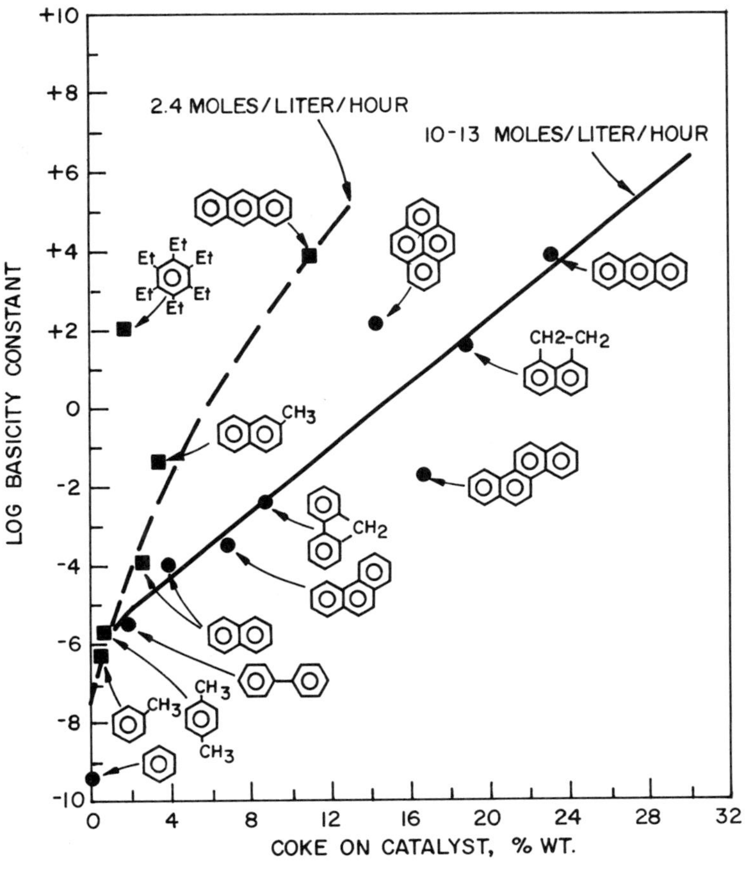

FIG. 48. Correlation between basicity of aromatic hydrocarbons and coke formation in catalytic cracking over amorphous silica-alumina-zirconia at 932° F; after Appleby, Gibson, and Good [302]. Note that the tendency to form coke generally increases with the degree of substitution of the aromatic nucleus. With hexaethylbenzene, coke formation may be less than expected, since its multiple substitution may actually decrease its ability to adsorb on the catalyst surface.

Rooney, and Kemball [230], in a study of propylene polymerization over silica-alumina at 140 to 392° F, deduced operation of a mechanism similar to that described above, with cyclization probably constituting the slow step in the redistribution process. The intracrystalline

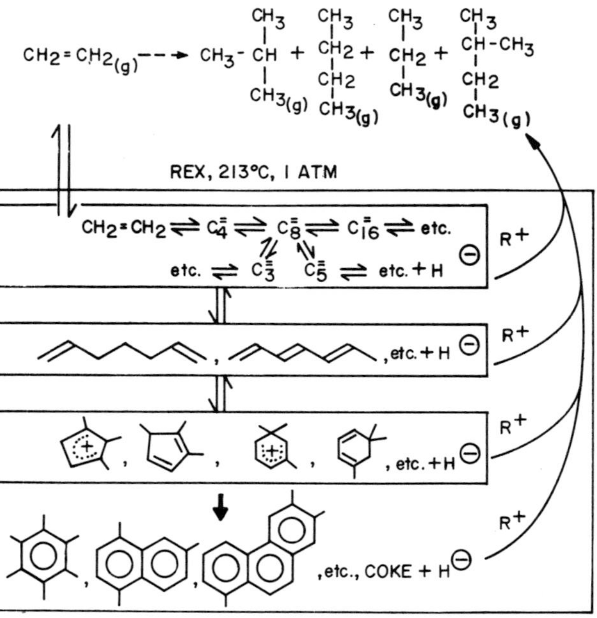

FIG. 49. Reaction of ethylene alone over rare earth-exchanged X-type zeolite at 415° F. Although initial building blocks were small and mobile, the size and rigidity of bulky aromatics synthesized within the zeolitic cavity became so great that escape by diffusion was impossible. The steric factors were also coupled with strong attraction of produced aromatics for the zeolite surface. Intracrystalline coke deposits and blockage of catalyst sites were the ultimate result. After Venuto [280].

polymerization of a low molecular weight olefin (1-hexene)—just prior to the onset of hydrogen redistribution—has been described [278, 285, 301] and is represented in Fig. 50. Van Hook and Emmett [104, 296] and Hightower and Emmett [303, 304] have reported an extensive secondary reaction of propylene labeled with ^{14}C during n-hexadecane cracking over silica-alumina at 700° F. Propylene participated extensively in the formation of aromatics—and saturates—generated during cracking of $C_{16}H_{34}$.

It is thus evident that, even at temperatures that are low relative to these commonly employed in catalytic cracking, hydrogen redistribution reactions of olefins can lead to aromatization and coke

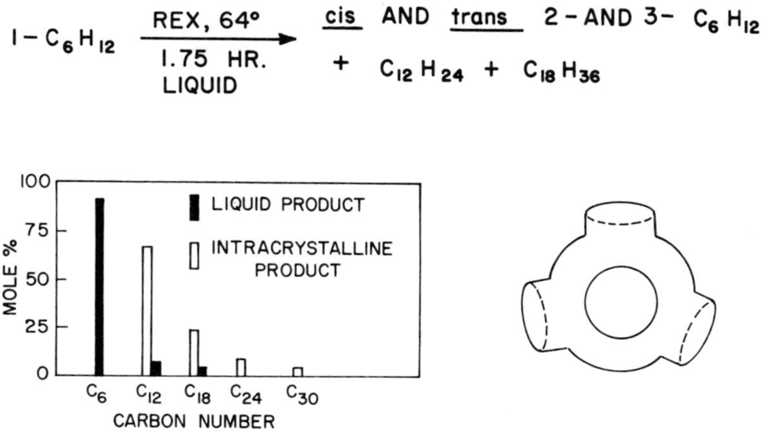

FIG. 50. Although no hydrogen transfer and aromatization had occurred at these relatively mild reaction conditions, accumulation within the pores of much higher molecular weight material than was present in the liquid product was observed. Such entrapment within the "bottle-like" pore (reverse molecular size selectivity) is a prelude to the formation of coke precursors. After Venuto [278].

formation. Indeed, in a study of the cracking behavior of a highly naphthenic distillate gas oil from which essentially all polycyclic and heterocyclic aromatics had been removed by extraction, John and Wojciechowski [300] concluded that the formation of coke derived from secondary reactions of product olefins.

D. Some Aspects of Gas Oil Cracking Kinetics

1. Correlations of Crackability with Compositional Properties and Reaction Severity

See Section VI-A for the related discussion on feedstock properties.

The term crackability indicates the relative propensity of a feed to undergo cracking at a given set of operating conditions [305]. Crackability shows a fairly good correlation with the amount of hydrogen contained in a gas oil. This is reflected in the data of Voltz, Nace, and Weekman [110] in Fig. 51, where the aromatic/naphthene ratio is inversely proportional to hydrogen content. Further, (catalytically) hydrogenated gas oils are more crackable than their

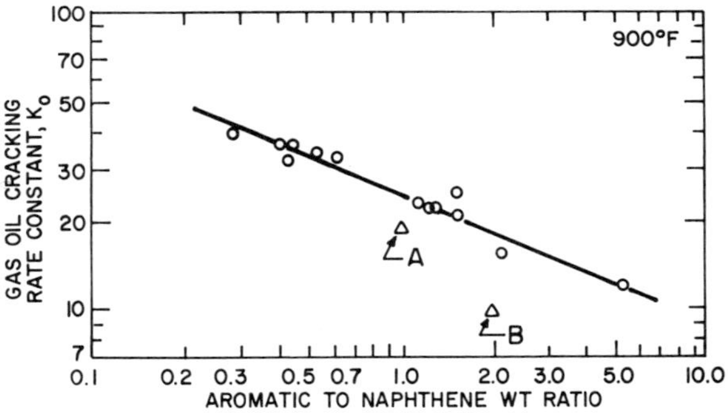

FIG. 51. Plot of gas oil cracking rate constant (k_0) vs aromatic/naphthene ratio for a series of virgin gas oils (o) in a fluidized dense bed reactor using zeolite catalyst; after Voltz, Nace, and Weekman [110]. The correlation is quite good except for points A (coker gas oil) and B (cycle stock). The deviation in B may relate to its considerably higher concentration of refractory aromatics; the distortion in A may arise from high concentrations of aromatics and olefins.

nonhydrotreated precursor stocks [39], indicating the creation of highly reactive hydroaromatic configurations upon chemical addition of hydrogen [123]. In fact, the cracking patterns of hydrotreated gas oils appear to be quite similar to those of good quality nonhydrotreated distillate gas oils [59, 99]. The rate of cracking can be significantly lowered by nitrogen poisoning (Fig. 52) [104]; this type of deactivation is severe, but activity can be restored by regeneration [11].

Crackability also varies with the extent of conversion since, as the gas oil mixture ascends the riser length, the concentration of unreacted, potentially convertible gas oil drops [66]. This is illustrated by the catalyst/oil (C/O) response (effect of C/O upon conversion) data shown in Fig. 53 [99]. At 900° F the response was +8 vol% conversion for doubling the C/O ratio in the 65 to 77 vol% conversion range; at 1000° F in the conversion range of 75 to 81 vol% the response for doubling C/O was +10 vol%. At 1100° F, even with the added thermal severity, the response for doubling C/O was only +2.2 vol%, since conversion had reached such high levels (85 to 90 vol%) that little incremental cracking could occur [99].

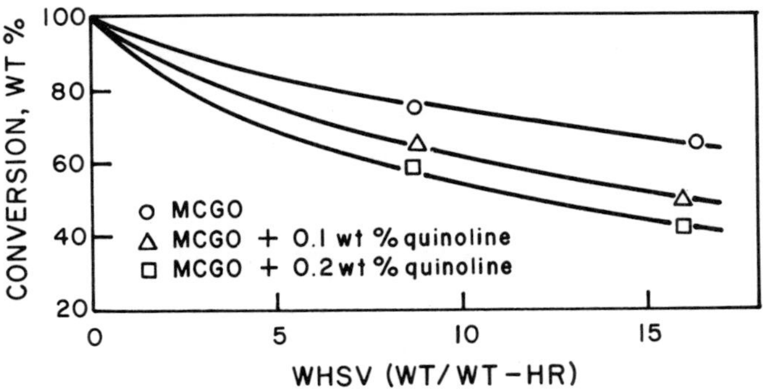

FIG. 52. Effect of basic nitrogen poisoning on conversion of Mid Continent gas oil (MCGO) in cracking over zeolite catalyst (fluidized dense bed reactor) at 1000° F and 1.25 min catalyst residence time. The conversion (time-averaged) drops sharply with increasing quinoline concentration in gas oil feed. After Voltz et al. [104].

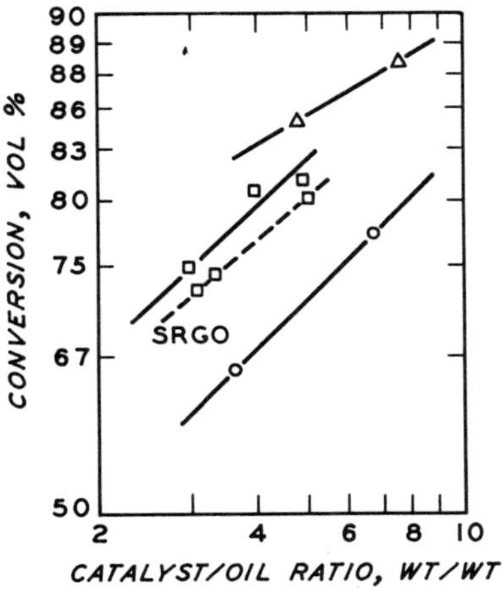

FIG. 53. Plot of conversion (vol%) vs catalyst/oil ratio (wt/wt) in riser cracking of hydrotreated heavy vacuum gas oil (HDT HVGO) at 4.5 sec residence time over fluid zeolite catalyst; after Owen, Snyder, and Venuto [99]. O, □, Δ = 900, 1000, and 1100° F, respectively. --- = data for a straight run gas oil (SRGO) of similar hydrogen content.

2. The Effect of Hydrocarbon Class upon Reactivity in Gas Oil Cracking: The Depth of Cracking Effect

In Section IX-D we emphasized that the different classes of pure hydrocarbons—paraffins, olefins, naphthenes, and aromatics—showed a wide range of cracking rates (see Fig. 38 for example). Thus, in the cracking of gas oil which contains a complex mixture of all classes of hydrocarbons, overall reactivity is itself a function of conversion [99, 241, 306]. Thus, in the resulting competition the more reactive components such as naphthenes and paraffins (particularly isoparaffins) are initially cracked away at very rapid rates [65, 99, 109, 122, 123]. As the reaction proceeds, however, there is a progressive enrichment in concentration of more refractory, chemically-resistant hydrocarbon species [99, 241], usually aromatics, which crack with greater difficulty [65, 123]. There is thus a decrease in reaction velocity.

This phenomenon is illustrated in Fig. 54. At the higher conversions, most of the aromatic fraction consists of substituent-free aromatic nuclei, the more reactive side chains having been removed in the first-pass operation [104]. It is thus not surprising that cycle oils (and coker stocks) have considerably lower rate constants for gas oil cracking than virgin gas oils [104, 109], as witnessed by Point B in Fig. 51. This depth of cracking effect also explains, in part, the C/O response patterns described above and shown in Fig. 53. There may also be some contribution to formation of refractory aromatics from hydrogen transfer reactions (Section IX-C-6).

3. Depth of Cracking and the Onset of Diffusion Limitations

In the discussion in Section IX-D-2, the dominant feature was competition based on chemical reactivity of competing hydrocarbon classes. It is also reasonable to expect that a certain fraction of the molecules present in a heavy gas oil might encounter some degree of diffusional resistance in gaining access to the intracrystalline pore volume of a molecular sieve catalyst [307, 308]. The "configurational regime" [197] (see Fig. 36) also applies to zeolites such as faujasite. Although these have relatively large pores (~10 Å), molecular size limitations, even to the exclusion or screening of larger molecules, will eventually occur.

Data from a study by Owen, Snyder, and Venuto [99] of FCC riser cracking of hydrotreated heavy vacuum gas oil may provide an example of a transition from a situation where chemical rates were

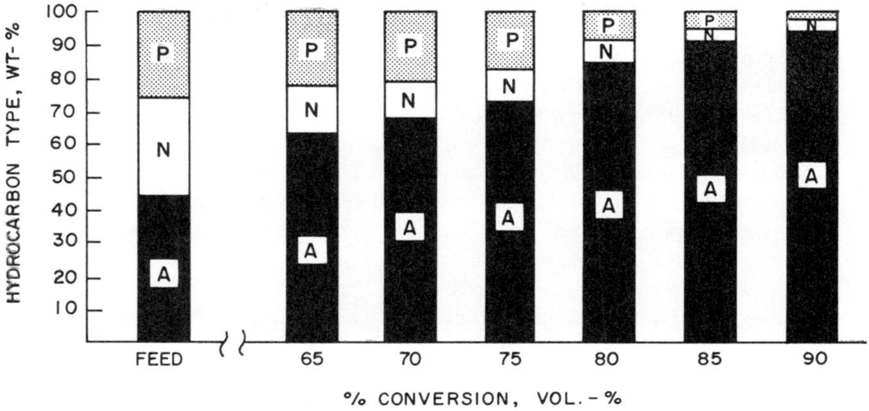

FIG. 54. Plot of hydrocarbon type (P, N, A) in cycle oils vs conversion level for FCC riser cracking of hydro-treated heavy vacuum gas oil at 900 to 1100° F over zeolitic catalyst; from data of Owen, Snyder and Venuto [99]. Cycle oils, which represent what did not crack in a given cracking pass, can be visualized as the reactant for the next increment of cracking; thus there is a progressive decline in crackability as the more reactive components are converted at higher severity.

controlling to a diffusion-controlled regime. At very high reaction severities and conversions (85 to 90 vol%), a flat C/O response (see Fig. 53, 1100° F line) and very low temperature response (apparent E_A 1.8 kcal/mole for the interval 1000 to 1100° F vs 12.9 kcal/mole for the interval 900 to 1000° F) were observed. The sharply lower activation energy was in the direction of values commonly encountered in severely diffusion-limited catalytic systems [309].

Analysis of the cycle oil from the highest conversion run (90 vol%, Fig. 54) showed the presence of 92.0 wt% aromatics, 4.8 wt% naphthenes, and only 3.3 wt% paraffins. There were at least two rings in all the aromatics and most of the naphthenes. Ninety-three percent of the aromatics contained two or more aromatic rings (including substituted naphthalenes, phenanthrenes, pyrenes—and heterocyclic sulfur analogs—with 13 to 17 carbon atoms and molecular weights of 170 to 225). The average size of molecules such as these is close to or even slightly larger than the 8 to 10 Å entry pores of faujasite. It is not difficult to imagine that such rigid, polarizable aromatics (with their tendency toward strong adsorption on polar surfaces) would have difficulty entering the intracrystalline pore-channel system, particularly if dimensions were already somewhat constricted by coke deposits. These data suggest operation of diffusion effects, or at least of extreme chemical inertness that mimics a diffusion limitation.

On the other hand, Wojciechowski [315], in considering the apparent drop-off in gas oil reactivity with rising temperature, suggests that increasing reaction order may be the determining kinetic factor.

Molecular Sieving/Shape Selectivity in Cracking over Faujasite Catalysts

Some studies of Nace [123] provide straightforward evidence for the existence of molecular size limits on cracking rates over REHX faujasite. Nace compared the rates of cracking of a series of naphthenes with increasingly bulky ring structure over zeolite and amorphous catalysts. (A comparison of the pore size distributions of these catalyst types is shown in Fig. 43.) As shown in Fig. 55, with silica-alumina the cracking rate constant increased continuously with size of naphthene reactant, with the C_{16}-naphthene showing over a three-fold rate enhancement over the paraffin of the same molecular weight. With REHX, however, the cracking rate constant increased with reactant size initially, then decreased sharply with the large 3- and 4-ring naphthenes. Correspondingly, the ratio $k_{REHX}/k_{SiO_2/Al_2O_3}$ dropped from a base value of 17 to a low of 2.4. This is clearly a case where the increasing size of the naphthene was associated with restricted intracrystalline diffusion—and consequently, cracking rate [123]. Similar studies with polynuclear aromatics showed irreversible adsorption and formation of coke [123].

4. Kinetic Rate Expressions and Reaction Networks

The Complexities Inherent in Heterogeneous Catalysis

As in any heterogeneous catalytic system, an involved sequence of steps must be followed [310] for the reactant to be transformed to the product. These include:

1. Transport of reactants from the bulk of the fluid phase to the external surface of the catalyst particle
2. Intraparticle diffusion of reactants through the pore system of the catalyst

CRACKING RATE CONSTANTS AT 900° F.
(2 min on-stream instantaneous value)

Reactant Hydrocarbon	SiO_2/Al_2O_3	REHX	Ratio $k_{REHX}/k_{SiO_2/Al_2O_3}$
$nC_{16}H_{34}$	60 (0.1%C)	1000 (1.4%C)	17
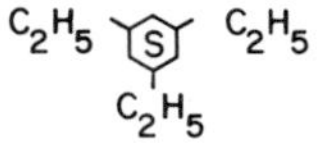	140 (0.4%C)	2370 (2.0%C)	17
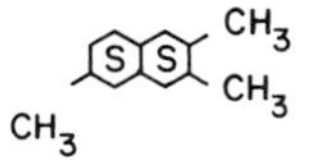	190 (0.2%C)	2420 (0.7%C)	13
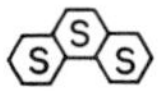	205 (0.2%C)	953 (1.0%C)	4.7
	210 (0.4%C)	513 (1.6%C)	2.4

FIG. 55. Comparison of cracking rates of n-hexadecane and a series of naphthenes of progressively larger ring structure over amorphous silica-alumina and zeolitic catalyst. After Nace [123].

3. Adsorption of reactants at the catalyst active sites
4. Chemical reaction on the surface of the catalyst
5. Desorption of products from surface
6. Diffusion of products from the interior pores to the outer surface of the catalyst particle
7. Transport of products from the fluid-solid interface, i.e., desorption to the bulk fluid stream

Although the rate of chemical reaction (Step 4) may be the slowest step and rate-controlling, there are many factors potentially complicating kinetic analysis in catalytic cracking. There is complex interplay between process variables and feed-product composition, and thermal/heat transfer factors will also affect the rate of reaction. A major problem in kinetic interpretation derives from the rapid, time-dependent decay of catalyst activity from coke formation that occurs during cracking [241, 253, 276, 284, 293, 306, 307, 311-318]. Moreover, there are complexities due to the multiplicity of reactants (and products) competing for entry to the catalyst pores, adsorption on the acidic surface and reaction, the great molar expansion during cracking, and the highly coupled networks of secondary reactions.

The Convergence toward Second-Order Kinetics in Gas Oil Cracking

Voge [11] has given an excellent assessment of basic kinetic principles in catalytic cracking with particular emphasis on reactions of pure compounds in fixed beds. Much past experience shows that data for cracking of simple pure hydrocarbons fit a first-order rate law [11, 41, 122, 312, 319] as in Eq. (15), where n_A is the number of moles of A present:

$$\frac{-dn_A}{dt} = k_1 n_A \tag{15}$$

Blanding [122], however, pursued the implications of the fact that the moles of product formed/mole of feed reacted increased considerably due to secondary reactions as the reaction proceeded. The data indicated that, although the basic reactions were first order because of dilution of reactant with cracked products, in a constant pressure system the reaction approached second order in terms of unconverted feed:

$$\frac{-dn_A}{dt} = k_2 P W_c \left(\frac{n_A}{n_A^0}\right)^2 \tag{16}$$

Here, P = total pressure, W_c = weight of catalyst, and $n_A/n_A°$ = fraction of reactants remaining (in a fixed bed, "piston-flow" reactor configuration). In practice, for the 30 to 80% conversion range, most feedstocks studied showed exponents ranging from 1.6 to 1.9. Blanding [122] further observed that the depth of cracking effect—the tendency of the most reactive species to crack first—also tended to increase the apparent reaction order above 1.

More recently, Weekman, Nace, Voltz, and associates [41, 104, 109, 110, 306], employing a different approach and working with fluid zeolite catalysts, also demonstrated that the kinetics of gas oil cracking could reasonably be represented as a (pseudo) second-order reaction coupled with catalyst decay. Although pure hydrocarbons crack according to the first-order rate law, the fact that the multiplicity of hydrocarbon constituents of gas oil crack at widely different rates allows the overall system to be well approximated by a second-order expression [41, 306]. The net effect of the depth of cracking phenomenon, despite the operation of many nonlinear effects, will be to increase the apparent order of the reaction since the average reactivity of unconverted feed molecules will decrease as conversion increases [306].

This has also been shown theoretically by Luss and co-workers [320-322] who studied methods of exact and approximate lumping for mixtures. In a situation in which many parallel first-order, irreversible reactions occur (reflecting the different reactivities of the diverse species), even though each will exhibit true first-order kinetics, the overall kinetics will lead to reaction orders higher than one if followed over a wide range of conversion. As observed by Satterfield [319], however, over a limited conversion range, first-order kinetics can often be exhibited by a group or lump of species of similar reactivity. Wojciechowski [315] cites an increase in reaction order (due to increase in feed refractoriness) in gas oil cracking at high temperatures.

Some Kinetic Networks and Models

One of the most comprehensive published kinetic models is that of Jacob et al. [311] which was evolved from refinements of the earlier model of Weekman and Nace [41] (the essence of which is represented in Fig. 41). The kinetic model is schematically shown in Fig. 56, and utilizes 10 major lumps to follow cracking of virgin gas oils and recycle stocks. Because petroleum fractions, even in relatively narrow boiling range cuts, contain thousands of individual molecular entities, lumping together of many constituents is necessary for efficient manipulation. In the model of Jacob et al. [311], the rate constants for the lumped species are invariant with respect to the composition of the charge stock.

As shown in Fig. 56, the key lumps comprise paraffins, naphthenes, bare aromatic rings, and aromatic substituent groups (side chains) in both light and heavy fuel oil cuts, as well as C_5-430° F gasoline (G-lump) and the aggregate H_2, H_2S, C_1 to C_4, coke (C-lump). Weight percent conversion is defined as the sum of the C- and G-lumps. The model incorporates terms to deal with the very rapid time-dependent catalyst deactivation that occurs during cracking of gas oil, as well as adsorption terms to account for nitrogen poisoning and heavy aromatic rings (the latter greatly influence the availability of catalyst acid sites). From this kinetic framework, predictions for (time-averaged) conversion, gasoline, light fuel oil, and recycle oil can be obtained as a function of process variables. The Voorhies relationship (Eq. 14) is used to predict coke yields.

Figure 57 depicts the kinetic scheme suggested by John and Wojciechowski [284] for cracking of an essentially aromatic-free neutral

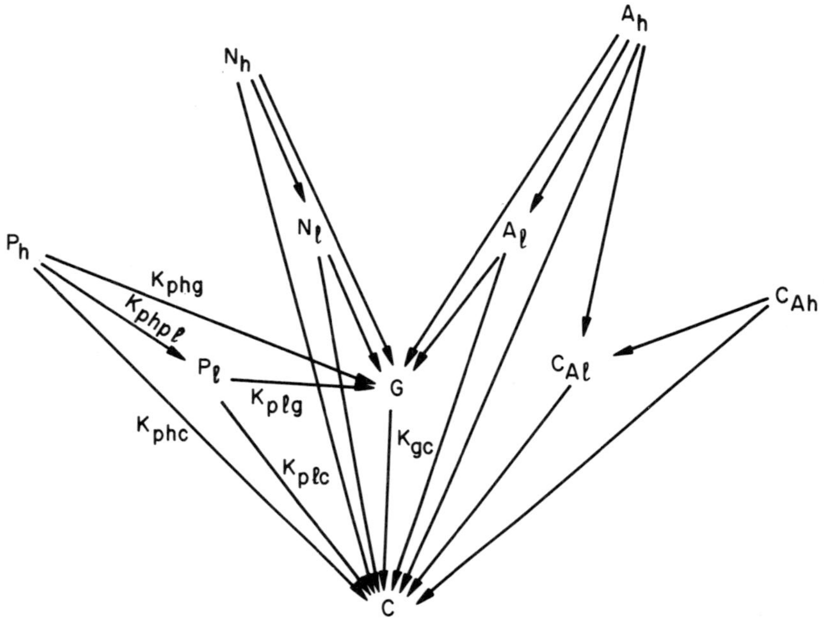

FIG. 56. Lumped kinetic model for fluid catalytic cracking; after Jacob et al. [311]. P_ℓ = wt% paraffinic molecules, 430 to 650° F; N_ℓ = wt% naphthenic molecules, 430 to 650° F; $C_{A\ell}$ = wt% carbon atoms among aromatic rings, 430 to 650° F; A_ℓ = wt% aromatic substituent groups, 430 to 650° F; P_h = wt% paraffinic molecules, 650° F+; N_h = wt% naphthenic molecules, 650° F+; C_{Ah} = wt% carbon atoms among aromatic rings, 650° F+; A_h = wt% aromatic substituent groups, 650° F+; G = G lump, C_5-430° F; C = C lump, C_1 to C_4 + coke; $C_{A\ell}$ + P_ℓ + N_ℓ + A_ℓ = LFO 430 to 650° F; C_{Ah} + P_h + N_h + A_h = HFO, 650° F+. Nomenclature is shown for paraffin rate constants.

distillate over a zeolite catalyst. As shown in Fig. 57, C_5^+ gasoline, butene, n-butane, and propylene are proposed to be the only primary products, with all others, including coke, arising solely from secondary reactions. Butene and gasoline are considered as unstable primary products which readily convert to form isobutane and some of the methane; propylene and n-butane are proposed as stable products, being generated from both feedstock and (primary product) butene and gasoline. Wojciechowski [253] also observes that as cracking temperature increases, the rate of formation of all secondary products increases, except coke.

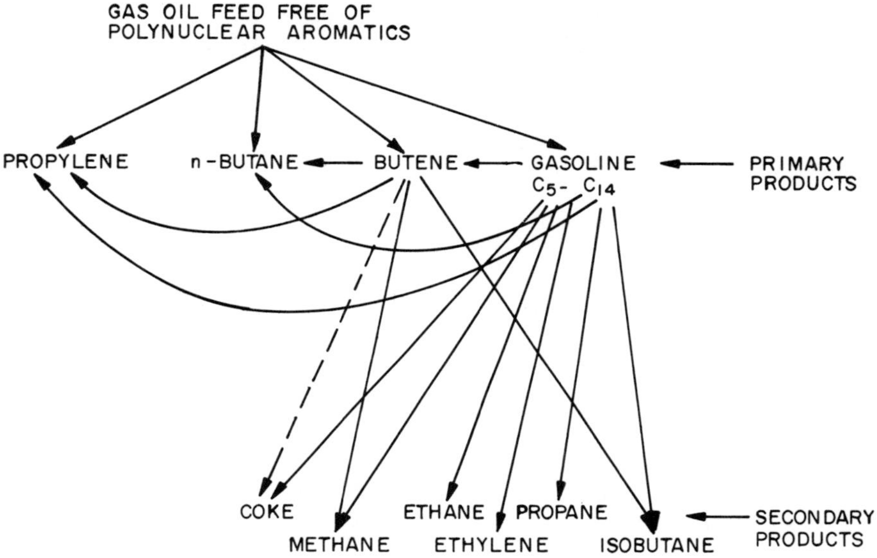

FIG. 57. Kinetic network of John and Wojciechowski [284] for cracking of gas oil from which almost all polycyclic and heterocyclic aromatics and metals had been removed by solvent extraction. The feed was highly naphthenic-paraffinic and the catalyst was LaY zeolite (fixed bed, 937° F).

Wollaston et al. [312] recently reported a cracking model that also took recognition of the fact that second-order rate expressions could be used to account for molar expansion and changing quality of (unconverted) feed. The second-order kinetics are reflected in the integrand x/(1 - x) in their expression correlating conversion with severity of reaction:

$$\frac{X}{1-X} = f(z_1, \ldots, z_m)\cdot(C/O)^n \cdot (WHSV)^{n-1} \cdot \exp^{(\Delta E/RT_{rX})} \tag{17}$$

where C/O = catalyst/oil ratio, lb/hr CCR/lb/hr oil rate
WHSV = weight hourly space velocity, wt of total feed/hr/wt of catalyst inventory in reaction zone, hr^{-1}
ΔE = activation energy for cracking, 25,000 Btu/lb-mole
R = gas constant, 1.987 Btu/lb-mole-°R
T_{rx} = kinetic average reaction temperature, °R
n = decay exponent, 0.65

x = volume-fraction conversion
f = function of feed quality, hydrocarbon partial pressure, intrinsic catalyst activity, and S_{C_R}, a severity factor reflecting the effective activity at various levels of carbon-on-regenerated catalyst, C_R (S_{C_R} is a function of process variables)

They also proposed an expression for coke yield that was consistent with Eq. (17), namely:

$$\text{COKE, Wt-\%} = g\,(z_1, \ldots\ldots, z_n)\cdot(C/O)^n\,(WHSV)^{n-1}\exp^{(\Delta E_c/RT_r X)} \tag{18}$$

where ΔE_C = activation energy for coke yield $\simeq$ 2500 Btu/lb-mole, and the remaining terms are similar to those in Eq. (17), particularly in the C/O and WHSV effect; n = 0.65 as above.
g = function similar to f in Eq. (17)

Acknowledgments

The authors would like to express their appreciation to Dr. C. J. Plank and Mr. E. J. Rosinski, pioneers in zeolite cracking catalysis, for many helpful discussions and perspectives in the cracking area over the years; and to Dr. P. B. Weisz, whose creativity and penetrating insight into the many dimensions of catalysis have been an inspiration and example.

REFERENCES

[1] J. J. McKetta, Chem. Eng. Prog., 69, 51 (1973).
[2] C. V. Kroeger, Hydrocarbon Process., 52(7), 57 (1973).
[3] E. A. Walters and E. M. Wewerka, J. Chem. Educ., 52, 282 (1975).
[4] L. G. Southard, International Petroleum Annual, 1971, Bureau of Mines, U.S. Dept. of Interior, March 1973.
[5] J. P. Dosher, Chem. Eng., 77(17), 96 (1970).
[6] A. Cantrell, Oil Gas J., 73(14), 96 (1975).
[7] D. H. Stormont, Ibid., 66(14), 104 (1968).
[8] L. F. Hatch, Hydrocarbon Process., 48(2), 77 (1969).
[9] E. J. Demmel, A. V. Perella, W. A. Stover, and J. P. Shambaugh, Proc. A.P.I. Div. Refining, 46, 165 (1966).

[10] M. Sittig, Pet. Refiner, 31(9), 263 (1952).
[11] H. H. Voge, in Catalysis, Vol. 6 (P. H. Emmett, ed.), Reinhold, New York, 1958, Chap. 5, p. 407.
[12] J. B. Pohlenz, Oil Gas J., 61(13), 124 (1963).
[13] A. G. Oblad, Ibid., 70(13), 84 (1972).
[14] W. L. Nelson, Petroleum Refining Engineering, McGraw-Hill, New York, 1941, p. 139.
[15] R. M. Milton, in Molecular Sieves, Soc. Chem. Ind., London, 1968, p. 199.
[16] R. M. Milton, U.S. Patent 2,882,244 (1959).
[17] C. J. Plank and E. J. Rosinski, Chem. Eng. Prog. Symp. Ser., 73(63), 26 (1967).
[18] C. J. Plank, E. J. Rosinski, and W. P. Hawthorne, Ind. Eng. Chem., Prod. Res. Dev., 3, 165 (1964).
[19] C. W. Strother, W. L. Vermilion, and A. J. Conner, Hydrocarbon Process., 51(5), 89 (1972).
[20] J. J. Blazek, Catalagram, 48, 1 (1975).
[21] B. Hitchon, Adv. Chem. Ser., 103, 30 (1971).
[22] U. Colombo, in Fundamental Aspects of Petroleum Geochemistry (B. Nagy and U. Colombo, eds.), Elsevier, Amsterdam, 1967, p. 331.
[23] G. W. Hodgson, Adv. Chem. Ser., 103, 1 (1971).
[24] G. W. Hodgson, B. L. Baker, and E. Peake, in Fundamental Aspects of Petroleum Geochemistry (B. Nagy and U. Colombo, eds.), Elsevier, Amsterdam, 1967, p. 177.
[25] J. E. Falk, Porphyrin and Metalloporphyrins, Elsevier, Amsterdam, 1964, p. 18.
[26] O. A. Larson and H. Beuther, Prepr. Div. Pet. Chem., Am. Chem. Soc., 11(2), B-95 (1966).
[27] J. M. Sugihara and J. F. Branthaver, Ibid., 21(3), 411 (1976).
[28] R. N. Cimbalo, R. L. Foster, and S. J. Wachtel, Oil Gas J., 70(20), 112 (1972).
[29] G. P. Masologites and L. H. Beckberger, Ibid., 71(47), 49 (1973).
[30] J. G. Bendoraitis, B. L. Brown, and L. S. Hepner, Proc. Sixth World Pet. Congr., 5, 13 (1963).
[31] E. D. Rossini and B. J. Mair, Adv. Chem. Ser., 5, 334 (1951).
[32] F. D. Rossini, J. Chem. Ed., 37, 554 (1960).
[33] J. S. Ball, W. E. Haines, and R. V. Helm, Proc. Fifth World Pet. Congr., 5, 175 (1959).
[34] M. A. Bestorgelf, in Fundamental Aspects of Petroleum Geochemistry (B. Nagy and U. Colombo, eds.), Elsevier, Amsterdam, 1967, p. 77.

[35] J. P. Dickie and T. F. Yen, Prepr., Dev. Pet. Chem., Am. Chem. Soc., 12(2), B-117 (1967).
[36] H. Tominaga and T. Kunugi, Int. Chem. Eng., 14(4), 753 (1974).
[37] P. A. Witherspoon and R. S. Winniford, in Fundamental Aspects of Petroleum Geochemistry (B. Nagy and U. Colombo, eds.), Elsevier, Amsterdam, 1967, p. 261.
[38] W. G. Appleby, J. W. Gibson, and G. M. Good, Ind. Eng. Chem. Process Des. Dev., 1(2), 102 (1962).
[39] C. L. Hemler and W. L. Vermilion, Oil Gas J., 71(45), 88 (1973).
[40] M. L. Whisman and F. G. Cotton, Ibid., 69(52), 11 (1971).
[41] V. W. Weekman, Jr. and D. M. Nace, AIChE J., 16(3), 397 (1970).
[42] W. L. Nelson, Oil Gas J., 70(32), 49 (1972).
[43] G. Costantinides and G. Arich, in Fundamental Aspects of Petroleum Geochemistry (B. Nagy and U. Colombo, eds.), Elsevier, Amsterdam, 1967, p. 109.
[44] G. A. Mills, E. R. Boedeker, and A. G. Oblad, J. Am. Chem. Soc., 72, 1554 (1950).
[45] W. F. Pansing and J. B. Malloy, Ind. Eng. Chem., Process Des. Dev., 4(2), 181 (1965).
[46] K. S. Tkhoang, B. V. Romanovskii, and K. V. Topchieva, Int. Chem. Eng., 8(1), 72 (1968).
[47] W. M. Haunschild, D. O. Chessmore, and B. G. Spars, Paper presented at the 79th National AIChE Meeting in Houston, Texas, March 1975.
[48] R. M. Koros, S. Bank, J. F. Hofmann, and M. I. Kay, Prepr. Div. Pet. Chem., Am. Chem. Soc., 12(4), B-165 (1967).
[49] O. A. Larson, Ibid., 12(4), B-123 (1967).
[50] G. P. Masologites and H. E. Jacobs, Oil Gas J., 73(21), 94 (1975).
[51] Anon., Hydrocarbon Process., 51(9), 189 (1972).
[52] Anon., Oil Gas J., 67(40), 91 (1969).
[53] L. R. Aalund, Ibid., 69(50), 67 (1971).
[54] R. C. Ewing, Ibid., 67(39), 60 (1969).
[55] R. W. Baker, J. J. Blazek, P. K. Maher, F. G. Ciapetta, and R. E. Evans, Ibid., 62(18), 78 (1964).
[56] H. N. Olson and K. E. Hutchinson, Ibid., 70(1), 39 (1972).
[57] L. R. Aalund, Ibid., 73(14), 86 (1975).
[58] R. O. Skamser, Ibid., 71(12), 71 (1973).
[59] R. E. Hildebrand, G. P. Huling, and G. F. Ondish, Ibid., 71(50), 112 (1973).

[60] E. K. Grigsby, E. W. Mills, and D. C. Collins, Hydrocarbon Process., 52(5), 133 (1973).
[61] C. W. Strother, W. L. Vermilion, and A. J. Conner, Oil Gas J., 70(20), 102 (1972).
[62] D. P. Bunn, Jr., G. F. Gruenke, H. B. Jones, D. C. Luessenhop, and D. J. Youngblood, Chem. Eng. Prog., 65(6), 88 (1969).
[63] A. L. Saxton and A. C. Worley, Oil Gas J., 68(20), 82 (1970).
[64] W. L. Pierce, R. P. Souther, T. G. Kaufman, and D. F. Ryan, Hydrocarbon Process., 51(5), 92 (1972).
[65] J. R. Murphy, Oil Gas J., 68(47), 72 (1970).
[66] W. L. Vermilion, Paper Presented at the Belgian Petroleum Institute, Antwerp, Belgium, November 1974.
[67] E. L. Whittington, J. R. Murphy, and I. H. Lutz, Oil Gas J., 70(44), 49 (1972).
[68] Anon., Chem. Week, 106(4), 31 (1970).
[69] J. W. Moorman, Oil Gas J., 53(36), 68 (1955).
[70] A. J. Conner, C. W. Strother, and W. L. Vermilion, U.S. Patent 3,785,962 (1974).
[71] W. L. Nelson, in Kirk-Othmer Encyclopedia of Chemical Technology, Vol. 15 (A. Standen, ed.), Wiley-Interscience, New York, 1966, p. 1.
[72] J. A. Finneran, J. R. Murphy, and E. L. Whittington, Oil Gas J., 72(2), 52 (1974).
[73] Anon., Ibid., 65(16), 161 (1967).
[74] E. L. Whittington, J. R. Murphy, and I. H. Lutz, Prepr. Div. Pet. Chem., Am. Chem. Soc., 17(3), B-66 (1972).
[75] G. S. Shnaider, Int. Chem. Eng., 13(3), 522 (1973).
[76] R. H. Ebel, Oil Gas J., 66(14), 116 (1968).
[77] M. C. Bryson, G. P. Huling, and W. E. Glausser, Ibid., 70(20), 97 (1972).
[78] J. H. Prescott, Chem. Eng., 81, 64 (1974).
[79] R. J. Shields, R. J. Fahrig, and C. J. Horecky, Oil Gas J., 70(22), 45 (1972).
[80] F. A. Zenz, in Kirk-Othmer Encyclopedia of Chemical Technology, Vol. 9 (A. Standen, ed.), Wiley-Interscience, New York, 1966, p. 398.
[81] L. Rheaume, W. E. Ritter, J. J. Blazek, and J. A. Montgomery, Oil Gas J., 74(21), 66 (1976).
[82] B. Kalen and F. A. Zenz, AIChE Symp. Ser., 70(137), 388 (1974).
[83] F. A. Zenz, Hydrocarbon Process., 54(5), 125 (1975).
[84] W. S. Letzsch, Catalagram, 47, 5 (1974).

[85] P. J. Celani, U.S. Patent 3,817,280 (1974).
[86] F. N. Fagan, J. E. Penick, and K. W. Schatz, U.S. Patent 3,904,548 (1975).
[87] S. M. Csicsery, H. F. Harnsberger, W. D. Hughes, and R. D. Searle, Ind. Eng. Chem., Process Des. Dev., 14(1), 93 (1975).
[88] A. P. Krueuding, Chem. Eng. Prog., 71, 56 (1975).
[89] S. S. Braun, Oil Gas J., 71(21), 128 (1973).
[90] J. B. Arant, Ibid., 72(7), 106 (1974).
[91] J. A. Montgomery, Catalagram, 37, 3 (1971).
[92] J. J. Blazek, Oil Gas J., 71(41), 65 (1973).
[93] R. M. Clinton and T. J. Puzniak, Ibid., 73(35), 126 (1975).
[94] J. A. Montgomery, Ibid., 70(50), 81 (1972).
[95] M. C. Bryson, G. P. Huling, and W. E. Glausser, Hydrocarbon Process., 51(5), 85 (1972).
[96] J. M. Phelps, Oil Gas J., 55(21), 8 (1957).
[97] L. J. McGovern, Prepr. Div. Pet. Chem., Am. Chem. Soc., 17(3), B-19 (1972).
[98] R. F. Anderson, Oil Gas J., 72(6), 78 (1974).
[99] H. Owen, P. W. Snyder, and P. B. Venuto, Proc. Sixth Int. Congr. Catal., 2, 1071 (1977).
[100] S. C. Eastwood, R. D. Drew, and F. D. Hartzell, Oil Gas J., 60(44), 152 (1962).
[101] S. C. Eastwood, C. J. Plank, and P. B. Weisz, Proc. Eighth World Pet. Congr., 4, 245 (1971).
[102] J. A. Montgomery, Catalagram, 39, 3 (1973).
[103] P. J. White, Oil Gas J., 66(21), 112 (1968).
[104] S. E. Voltz, D. M. Nace, S. M. Jacob, and V. W. Weekman, Jr., Ind. Eng. Chem., Process Des. Dev., 11(2), 261 (1972).
[105] S. C. Eastwood, R. A. Sailor, and A. B. Schwartz, Proc. Seventh World Pet. Congr., 4, 89 (1967).
[106] R. W. Baker, P. K. Maher, and J. J. Blazek, Hydrocarbon Process., 17(2), 125 (1968).
[107] J. J. Blazek, Oil Gas J., 69(45), 66 (1971).
[108] J. G. Ditman, Ibid., 72(7), 84 (1974).
[109] D. M. Nace, S. E. Voltz, and V. W. Weekman, Jr., Ind. Eng. Chem., Process Des. Dev., 10(4), 530 (1971).
[110] S. E. Voltz, D. M. Nace, and V. W. Weekman, Jr., Ibid., 10(4), 538 (1971).
[111] D. E. Nicholson, Ind. Eng. Chem., 47(6), 1216 (1955).
[112] B. Gross, D. M. Nace, and S. E. Voltz, Ind. Eng. Chem., Process Des. Dev., 13(3), 199 (1974).

[113] R. N. Fleck and C. G. Wight, U.S. Patent 2,962,435 (1968).
[114] E. J. Demmel and H. Owen, U.S. Patent 3,791,962 (1974).
[115] N. Y. Chen, Proc. Fifth Int. Congr. Catal., 2(97), 1343 (1972).
[116] Anon., Pet. Times, 80(2035), 21 (1976).
[117] W. L. Nelson, Oil Gas J., 59(43), 143 (1961).
[118] J. R. Murphy and M. Soudek, Ibid., 75(3), 70 (1977).
[119] J. Van Pool, U.S. Patent 3,542,668 (1970).
[120] W. L. Nelson, Oil Gas J., 71(50), 130 (1973).
[121] Anon., Ibid., 70(14), 30 (1972).
[122] F. H. Blanding, Ind. Eng. Chem., 45(6), 1186 (1953).
[123] D. M. Nace, Ind. Eng. Chem., Prod. Res. Dev., 9(2), 203 (1970).
[124] R. W. Baker and J. J. Blazek, Proc. A.P.I. Div. Refining, 46, 274 (1966).
[125] J. S. Magee, J. J. Blazek, and R. E. Ritter, Oil Gas J., 71(30), 48 (1973).
[126] E. G. Wollaston, W. L. Forsythe, and I. A. Vasalos, Proc. A. P. I. Div. Refining, 51, 12 (1971).
[127] R. W. Baker, J. J. Blazek, and P. K. Maher, Oil Gas J., 66(14), 110 (1968).
[128] R. W. Baker and F. G. Ciapetta, Proc. A.P.I. Div. Refining, 44(III), 452 (1964).
[129] S. J. Wachtel, L. A. Baillie, R. L. Foster, and H. E. Jacobs, Prepr. Div. Pet. Chem., Am. Chem. Soc., 16(3), A-55 (1971).
[130] R. E. Ritter, J. J. Blazek, and D. N. Wallace, Oil Gas J., 72(41), 99 (1974).
[131] K. M. Elliott and S. C. Eastwood, Proc. A.P.I. Div. Refining, 42(III), 272 (1962).
[132] M. D. Edgar, Oil Gas J., 72(22), 166 (1974).
[133] D. V. Porchey and D. J. Royer, U.S. Patent 3,803,259 (1974).
[134] P. T. Atteridge and J. Humble, Pet. Refiner, 42(4), 167 (1963).
[135] A. V. Agafonov, I. E. Gelms, and E. I. Rabinovich, Int. Chem. Eng., 1(1), 1 (1961).
[136] Anon., Chem. Eng., 83(25), 86 (1976).
[137] A. Billon, J. P. Peries, E. Fehr, and E. Lorenz, Oil Gas J., 75(4), 43 (1977).
[138] J. B. Pohlenz, Ibid., 68(32), 158 (1970).
[139] D. H. Stormont, Ibid., 66(14), 114 (1968).

[140] J. F. Jones, P. L. Barrick, and J. H. Blake, Prepr. Div. Pet. Chem., Am. Chem. Soc., 11(2), C-79 (1966).
[141] J. A. Montgomery, Catalagram, 38, 3 (1971).
[142] W. F. Arey, Jr. and L. Kronenberger, Oil Gas J., 66(20), 133 (1969).
[143] D. H. Stormont, Ibid., 66(14), 104 (1968).
[144] D. H. Stormont, Ibid., 66(8), 40 (1968).
[145] P. J. White, J. F. Jones, and R. T. Eddinger, Hydrocarbon Process., 47(12), 97 (1968).
[146] J. R. Murphy, M. R. Smith, and C. H. Viens, Oil Gas J., 68(23), 108 (1970).
[147] D. C. McCulloch, Ibid., 73(29), 53 (1975).
[148] Anon., Ibid., 70(9), 43 (1972).
[149] Anon., Ibid., 74(20), 115 (1976).
[150] R. E. Ritter, Ibid., 73(36), 41 (1975).
[151] P. B. Weisz, U.S. Patent 3,413,212 (1968).
[152] G. P. Huling, J. D. McKinney, and T. C. Readal, Oil Gas J., 73(20), 73 (1975).
[153] D. Milstein, R. G. Graven, C. W. Streed, and P. C. Fusilier, Ibid., 74(29), 138 (1976).
[154] E. M. Gladrow, R. W. Krebs, and C. N. Kimber, Jr., Ind. Eng. Chem., 45(1), 142 (1953).
[155] J. J. Blazek, F. E. Ivey, Jr., and R. W. Baker, Proc. A.P.I. Div. Refining, 42(III), 277 (1962).
[156] C. J. Plank and E. J. Rosinski, U.S. Patent 3,140,249 (1964).
[157] C. J. Plank and E. J. Rosinski, U.S. Patent 3,140,253 (1964).
[158] C. J. Plank and E. J. Rosinski, U.S. Patent 3,210,267 (1965).
[159] C. J. Plank and E. J. Rosinski, U.S. Patent 3,271,418 (1966).
[160] C. J. Plank and E. J. Rosinski, U.S. Patent 3,436,357 (1969).
[161] C. J. Plank and E. J. Rosinski, U.S. Patent 3,459,680 (1969).
[162] W. F. Pansing, J. Phys. Chem., 69, 392 (1965).
[163] C. J. Plank and D. M. Nace, Ind. Eng. Chem., 47, 2374 (1955).
[164] P. Andreu, G. Martin, and H. Noller, J. Catal., 21, 255 (1971).
[165] R. C. Archibald, N. C. May, and B. S. Greensfelder, Ind. Eng. Chem., 44(8), 1811 (1952).
[166] A. Voorhies, Jr., Ibid., 37(4), 318 (1945).
[167] P. B. Weisz and A. B. Schwartz, J. Catal., 1, 399 (1962).
[168] P. B. Weisz, Science, 179, 433 (1973).
[169] H. S. Bloch and C. L. Thomas, J. Am. Chem. Soc., 66, 1589 (1944).

[170] R. W. Blue and C. J. Engle, Ind. Eng. Chem., 43(2), 494 (1951).
[171] G. Egloff, J. C. Morrell, C. L. Thomas, and H. S. Bloch, J. Am. Chem. Soc., 61, 3571 (1939).
[172] E. M. Gladrow and C. N. Kimberlin, Jr., Paper Presented at the 138th Am. Chem. Soc. Meeting, New York, September 1960.
[173] C. J. Plank, D. J. Sibbett, and R. B. Smith, Ind. Eng. Chem., 49(4), 742 (1957).
[174] R. M. Dobres, L. Rheaume, and F. G. Ciapetta, Ind. Eng. Chem., Prod. Res. Dev., 5(2), 174 (1966).
[175] C. L. Thomas, J. Am. Chem. Soc., 66, 1586 (1944).
[176] H. H. Voge, G. M. Good, and B. S. Greensfelder, Ind. Eng. Chem., 38(10), 1033 (1946).
[177] B. S. Greensfelder and H. H. Voge, Ibid., 37(6), 514 (1945).
[178] B. S. Greensfelder and H. H. Voge, Ibid., 37(11), 1038 (1945).
[179] G. M. Good, H. H. Voge, and B. S. Greensfelder, Ibid., 39(8), 1032 (1947).
[180] B. S. Greensfelder, H. H. Voge, and G. M. Good, Ibid., 41(11), 2573 (1949).
[181] R. L. Koffler, Paper Presented at the AIChE Materials Conference in Philadelphia, March 1968.
[182] J. S. Magee and J. J. Blazek, in American Chemical Society Monograph No. 171, Zeolite Chemistry and Catalysis (J. A. Rabo, ed.), American Chemical Society, Washington, D.C., 1976, p. 615.
[183] D. W. Breck, Zeolite Molecular Sieves, Wiley, New York, 1974.
[184] E. M. Flanigen, Adv. Chem. Ser., 121, 119 (1973).
[185] W. F. Arey, Jr., Oil Gas J., 68(3), 64 (1970).
[186] H. S. Sherry, in Ion Exchange, Vol. 2 (J. A. Marinsky, ed.), Dekker, New York, 1969, Chap. 3.
[187] H. S. Sherry, Adv. Chem. Ser., 101, 350 (1971).
[188] A. B. Schwartz, U.S. Patent 2,900,349 (1959).
[189] J. S. Magee, Catalagram, 41, 3 (1972).
[190] C. L. Thomas and D. S. Barmby, J. Catal., 12, 341 (1968).
[191] F. G. Ciapetta, Chim. Ind., 51(11), 1173 (1969).
[192] W. M. Meier and D. H. Olson, Adv. Chem. Ser., 102, 155 (1971).
[193] J. A. Rabo and M. L. Poutsma, Ibid., 102, 284 (1971).
[194] J. V. Smith, Ibid., 102, 171 (1971).

[195] J. Turkevich and Y. Ono, Ibid., 102, 315 (1971).
[196] P. B. Venuto and P. S. Landis, Adv. Catal., 18, 259 (1968).
[197] P. B. Weisz, Chem. Technol., 1973, 498.
[198] D. H. Olson, G. T. Kokotailo, and J. F. Charnell, J. Colloid Interface Sci., 28, 305 (1968).
[199] J. Scherzer, J. L. Bass, and F. D. Hunter, J. Phys. Chem., 79, 1194 (1975).
[200] B. J. Mair and M. Shamaiengar, Anal. Chem., 30, 276 (1958).
[201] C. L. Thomas and D. S. Barmby, J. Catal., 12, 341 (1968).
[202] P. B. Venuto and L. A. Hamilton, Ind. Eng. Chem., Prod. Res. Dev., 6, 190 (1967).
[203] N. Y. Chen and P. B. Weisz, Chem. Eng. Prog. Symp. Ser., 63(73), 86 (1967).
[204] P. B. Weisz, V. J. Frilette, R. W. Maatman, and E. B. Mower, J. Catal., 1, 307 (1962).
[205] J. A. Rabo and P. H. Kasai, in Progress in Solid State Chemistry (J. O. McCaldin and G. A. Somorjai, eds.), Pergamon, New York, 1975, p. 1.
[206] H. S. Sherry, J. Colloid Interface Sci., 28, 288 (1968).
[207] F. D. Hunter and J. Scherzer, J. Catal., 20, 246 (1971).
[208] J. Ward, Adv. Chem. Ser., 101, 380 (1970).
[209] R. Beaumont, D. Barthomeuf, and Y. Trambouze, Ibid., 102, 327 (1971).
[210] L. Moscou, Ibid., 102, 337 (1971).
[211] N. W. Cant and W. K. Hall, J. Catal., 25, 161 (1972).
[212] R. M. Barrer, F. W. Bultitude, and J. W. Sutherland, Trans. Faraday Soc., 53, 1111 (1957).
[213] P. B. Weisz and J. N. Miale, J. Catal., 4, 527 (1965).
[214] A. G. Oblad, Oil Gas J., 70(13), 84 (1972).
[215] C. V. McDaniel and P. K. Maher, Soc. Chem. Ind. (London), 1968, 186.
[216] C. D. Prater, Chem. Eng. Sci., 8, 284 (1958).
[217] J. Wei, Ibid., 12, 1171 (1966).
[218] P. B. Weisz and R. D. Goodwin, J. Catal., 2(5), 397 (1963).
[219] J. C. M. Lee and D. Luss, Ind. Eng. Chem., Fundam., 8, 596 (1969).
[220] P. B. Weisz, Ibid., 8(2), 325 (1969).
[221] H. R. Grane, J. E. Conner, and G. P. Masologites, Pet. Refiner, 40(5), 168 (1961).
[222] E. C. Gossett, Paper Presented at 52nd National AIChE Meeting in San Francisco, December 1959.
[223] J. E. Connor, J. J. Rothrock, E. R. Birkhimer, and L. N. Leum, Ind. Eng. Chem., 49(5), 278 (1957).

[224] M. Mocki, Chem. Abstr., 74, 128480 (1970).
[225] B. F. Morozov, R. A. Danilova, R. M. Masagutov, and G. A. Berg, Ind. Chem. Eng., 13(2), 324 (1973).
[226] J. J. Blazek, Catalagram, 42, 3 (1973).
[227] W. S. Letzsch, R. E. Ritter, and D. E. W. Vaughan, Oil Gas J., 74(4), 130 (1976).
[228] D. Kunii and O. Levenspiel, Fluidization Engineering, Wiley, New York, 1969.
[229] K. W. Schatz, H. Owen, and P. W. Snyder, Paper Presented at AIChE Meeting in Munich, September 1974.
[230] F. E. Shephard, J. J. Rooney, and C. Kemball, J. Catal., 1, 379 (1962).
[231] J. S. Bass, Catalagram, 49, 3 (1975).
[232] M. D. Edgar, Oil Gas J., 71(39), 110 (1973).
[233] F. G. Ciapetta and D. S. Henderson, Ibid., 65(42), 88 (1967).
[234] J. N. Miale, N. Y. Chen, and P. B. Weisz, J. Catal., 6, 278 (1966).
[235] W. K. Hall and D. S. MacIver, Ind. Eng. Chem., 52(5), 421 (1960).
[236] N. Y. Chen and S. J. Lucki, Ind. Eng. Chem., Process Des. Dev., 10(1), 71 (1971).
[237] D. M. Nace, Ind. Eng. Chem., Prod. Res. Dev., 8(1), 24 (1969).
[238] M. E. Conn and G. C. Connolly, Ind. Eng. Chem., 39(9), 1138 (1947).
[239] W. R. Gustafson, Ind. Eng. Chem., Process Des. Dev., 11(4), 507 (1972).
[240] L. Rheaume, W. E. Ritter, J. J. Blazek, and J. A. Montgomery, Oil Gas J., 74(20), 103 (1976).
[241] I. Wang, Thesis, University of Utah, 1974.
[242] J. J. Carberry and J. B. Butt, Catal. Rev., 10(2), 221 (1974).
[243] L. K. Doraiswamy and D. G. Tajbl, Ibid., 10(2), 177 (1974).
[244] V. W. Weekman, Jr., AIChE J., 20(5), 833 (1974).
[245] O. A. Wunderlich and F. E. Ivey, Jr., Oil Gas J., 56(3), 121 (1958).
[246] D. M. Himmelblau, Basic Principles and Calculations in Chemical Engineering, 3rd ed., Prentice-Hall, Englewood Cliffs, New Jersey, 1974, p. 288.
[247] D. I. Orochko and G. N. Chernakova, Int. Chem. Eng., 9(2), 219 (1969).
[248] P. B. Weisz, Ann. Rev. Phys. Chem., 21, 175 (1970).
[249] Anon., Oil Gas J., 71(14), 32 (1973).
[250] C. R. Olsen and M. J. Sterba, Chem. Eng. Prog., 45, 692 (1949).

[251] E. V. Murphree, C. L. Brown, H. G. M. Fischer, E. J. Gohr, and W. J. Sweeney, Ind. Eng. Chem., 35(7), 523 (1943).
[252] R. F. Anderson, Oil Gas J., 72(6), 78 (1974).
[253] T. M. John and B. W. Wojciechowski, J. Catal., 37, 348 (1975).
[254] D. M. Nace, Ind. Eng. Chem., Prod. Res. Dev., 8(1), 31 (1969).
[255] Anon., Catalagram, 51, 6 (1976).
[256] L. Aalund, Oil Gas J., 74(48), 60 (1976).
[257] W. L. Nelson, Ibid., 72(15), 66 (1974).
[258] O. A. Hougen, K. M. Watson, and R. A. Ragatz, Chemical Process Principles, 2nd ed., Part 1, Wiley, New York, 1954, Chap. 12, p. 502.
[259] H. V. Stuart, Catalagram, 45, 12 (1974).
[260] E. T. Habib, Jr., H. Owen, P. W. Snyder, C. W. Streed, and P. B. Venuto, Ind. Eng. Chem., Prod. Res. Dev., 16, 291 (1977).
[261] R. R. Edison, J. O. Siemssen, and G. P. Masologites, Oil Gas J., 74(51), 54 (1976).
[262] Anon., Ibid., 74(15), 83 (1976).
[263] E. C. French, Ibid., 74(15), 62 (1976).
[264] G. P. Huling, J. D. McKinney, and T. C. Readal, Ibid., 73(20), 73 (1975).
[265] J. H. Somers, R. J. Garbe, C. E. Fett, and T. M. Baines, Prepr. Div. Pet. Chem., Am. Chem. Soc., 21(4), 784 (1976).
[266] P. B. Weisz and R. D. Goodwin, J. Catal., 6, 227 (1966).
[267] P. B. Weisz, Ibid., 6, 425 (1966).
[268] J. R. Arthur, Trans. Faraday Soc., 47, 164 (1951).
[269] H. J. Lovink, in Ketjen Fluid Cracking Catalysts—Akzo Chemie (L. Moscou, ed.), Akzo Chemie, Amersfoort, 1975, Chap. 8.
[270] W. Lee, Ind. Eng. Chem., Process Des. Dev., 9, 154 (1970).
[271] Anon,, Chem. Eng., 79, 54 (1972).
[272] S. Gussow, G. W. Higginson, and I. A. Schwint, Hydrocarbon Process., 51(6), 116 (1972).
[273] R. E. Evans and C. O. McKinney, U.S. Patent 3,817,872 (1974).
[274] V. Haensel, Adv. Catal., 3, 179 (1951).
[275] D. R. Stull, E. F. Westrum, Jr., and G. C. Sinke, The Chemical Thermodynamics of Organic Compounds, Wiley, New York, 1969.
[276] M. L. Poutsma, in American Chemical Society Monograph No. 171, Zeolite Chemistry and Catalysis (J. A. Rabo, ed.), American Chemical Society, Washington, D.C., 1976, p. 529.

[277] J. C. Dart and A. G. Oblad, Chem. Eng. Prog., 45, 110 (1945).
[278] P. B. Venuto, Adv. Chem. Ser., 102, 260 (1971).
[279] S. M. Csicsery, American Chemical Society Monograph No. 171, Zeolite Chemistry and Catalysis (J. A. Rabo, ed.), American Chemical Society, Washington, D.C., 1976.
[280] P. B. Venuto, in Catalysis in Organic Synthesis, 1977 (G. V. Smith, ed.), Academic, New York, 1977.
[281] J. A. Rabo and M. L. Poutsma, Adv. Chem. Ser., 102, 284 (1971).
[282] B. S. Greensfelder and H. H. Voge, Ind. Eng. Chem., 37(10), 983 (1945).
[283] B. C. Gates, J. R. Katzer, and G. C. A. Schuit, Chemistry of Catalytic Processes, Academic, New York, 1977.
[284] T. M. John and B. W. Wojciechowski, J. Catal., 37, 240 (1975).
[285] P. B. Venuto, Chem. Technol., 1, 215 (1971).
[286] J. E. Germain, Catalytic Conversions of Hydrocarbons, Academic, London, 1969.
[287] W. F. Taylor, J. H. Sinfelt, and D. J. C. Yates, J. Phys. Chem., 69, 3857 (1965).
[288] P. N. Galich, V. S. Gutyrya, I. E. Neimark, Yu. P. Egorov, V. G. Il'in, I. T. Golubchenko, and V. W. Frolova, Neftekhim., Akad. Nauk Ukr. SSR, Inst. Khim. Vysokomol. Soedin., 1964, 13.
[289] D. W. McKee, J. Am. Chem. Soc., 84, 1109 (1962).
[290] J. Sinfelt, Prog. Solid State Chem., 10(2), 55 (1975).
[291] R. J. Mikovsky and J. F. Marshall, J. Catal., 44, 170 (1976).
[292] L. Moscou and R. Moné, Ibid., 30, 417 (1973).
[293] V. W. Weekman, Ind. Eng. Chem., Process Des. Dev., 8(3), 385 (1969).
[294] W. A. Van Hook and P. H. Emmett, J. Am. Chem. Soc., 85, 697 (1963).
[295] R. C. Haldeman and M. C. Botty, J. Phys. Chem., 63, 489 (1959).
[296] W. A. Van Hook and P. H. Emmett, J. Am. Chem. Soc., 84, 4421 (1962).
[297] J. L. Bordley, Jr. and P. H. Emmett, J. Catal., 42, 367 (1976).
[298] M. E. Levinter, G. M. Panchenkov, and M. A. Tanatarov, Int. Chem. Eng., 7(1), 23 (1967).
[299] C. L. Thomas and E. J. McNelis, Proc. Seventh World Pet. Congr., 1B, 161 (1967).

[300] T. M. John, R. A. Pachovsky, and B. W. Wojciechowski, Chem. Reaction Eng., 2, 422 (1974).

[301] P. B. Venuto, L. A. Hamilton, and P. S. Landis, J. Catal., 5, 484 (1966).

[302] W. G. Appleby, J. W. Gibson, and G. M. Good, Ind. Eng. Chem., Process Des. Dev., 1(2), 102 (1962).

[303] J. W. Hightower and P. H. Emmett, J. Am. Chem. Soc., 87, 939 (1965).

[304] J. W. Hightower and P. H. Emmett, Proc. Third Int. Congr. Catal., 1, 688 (1965).

[305] C. L. Thomas, Ind. Eng. Chem., 41, 2564 (1949).

[306] V. W. Weekman, Jr., Ind. Eng. Chem., Process Des. Dev., 7(1), 91 (1968).

[307] R. A. Pachovsky, D. A. Best, and B. W. Wojciechowski, Ibid., 12(3), 254 (1973).

[308] R. A. Pachovsky and B. W. Wojciechowski, AIChE J., 19(6), 1121 (1973).

[309] C. N. Satterfield, Mass Transfer in Heterogeneous Catalysis, M.I.T. Press, Cambridge, Massachusetts, 1970, Chap. 1, p. 4.

[310] J. M. Smith, Chemical Engineering Kinetics, 2nd ed., McGraw-Hill, New York, 1970, p. 274.

[311] S. M. Jacob, B. Gross, S. E. Voltz, and V. W. Weekman, Jr., AIChE J., 22, 701 (1976).

[312] E. G. Wollaston, W. J. Haflin, W. D. Ford, and G. J. D'Souza, Hydrocarbon Process., 54(19), 93 (1975).

[313] B. W. Wojciechowski, Can. J. Chem. Eng., 46, 48 (1968).

[314] B. W. Wojciechowski, J. A. Juusola, and J. Downie, Ibid., 47, 338 (1969).

[315] B. W. Wojciechowski, Catal. Rev., 9(1), 79 (1974).

[316] R. A. Pachovsky and B. W. Wojciechowski, Can. J. Chem. Eng., 52, 625 (1974).

[317] R. A. Pachovsky and B. W. Wojciechowski, Ibid., 49, 365 (1971).

[318] R. A. Pachovsky, T. M. John, and B. W. Wojciechowski, AIChE J., 19(4), 802 (1973).

[319] C. N. Satterfield, Ibid., 21(2), 209 (1975).

[320] P. Hutchinson and D. Luss, Chem. Eng. J., 1, 129 (1970).

[321] S. Golikeri and D. Luss, AIChE J., 18(2), 277 (1972).

[322] D. Luss and P. Hutchinson, Chem. Eng. J., 2, 172 (1971).

[323] V. J. Frilette, P. B. Weisz, and R. L. Golden, J. Catal., 1, 301 (1962).

INDEX